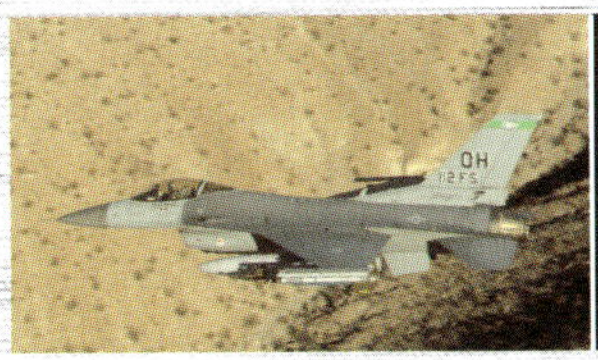

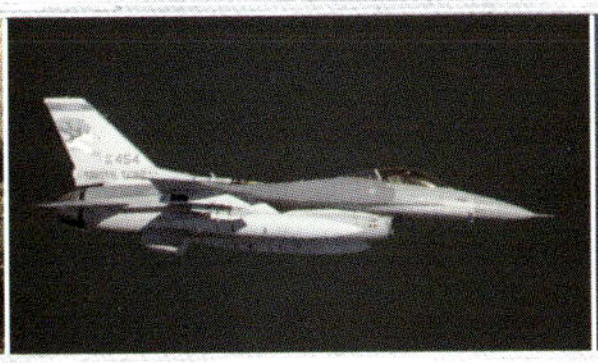

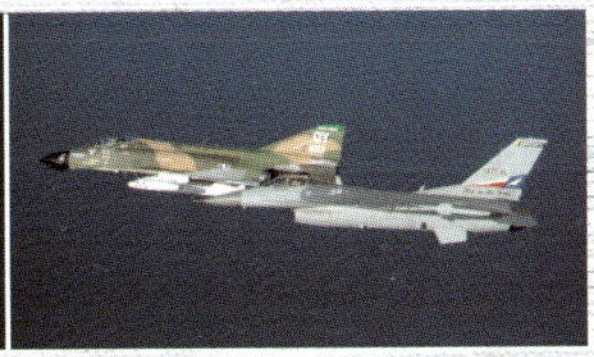

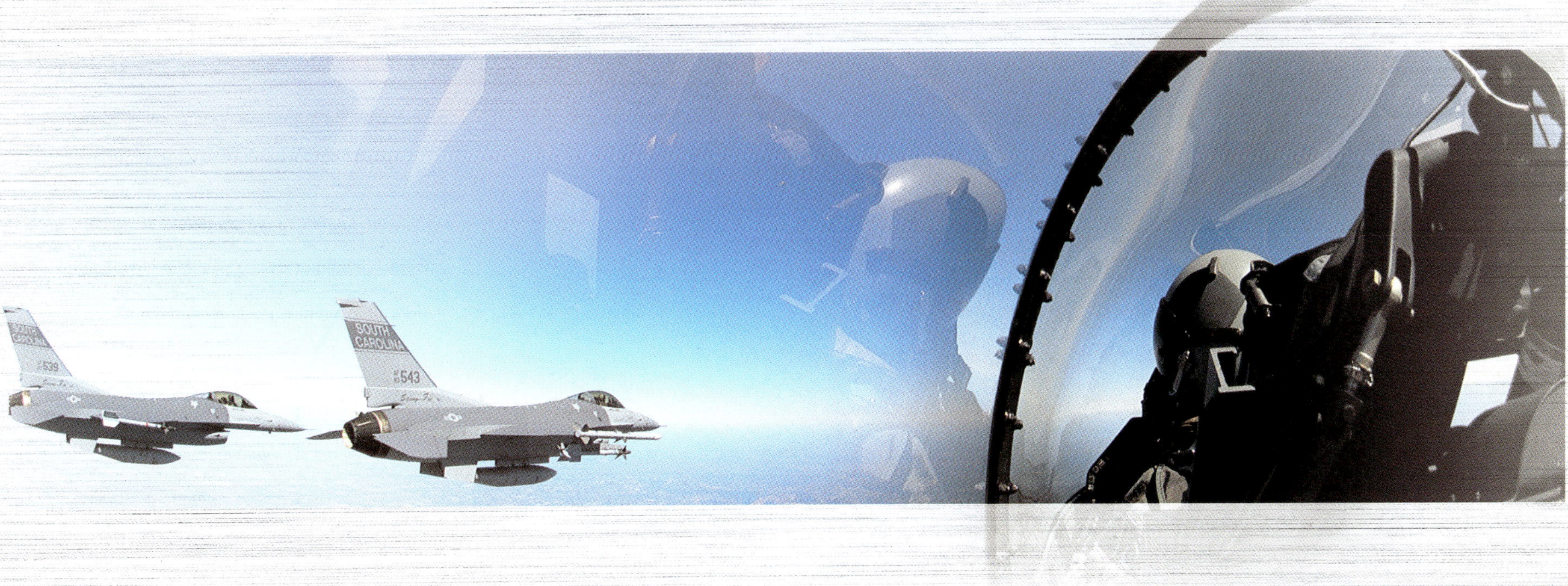

VIPERS IN THE SKY

GIAMPAOLO AGOSTINELLI

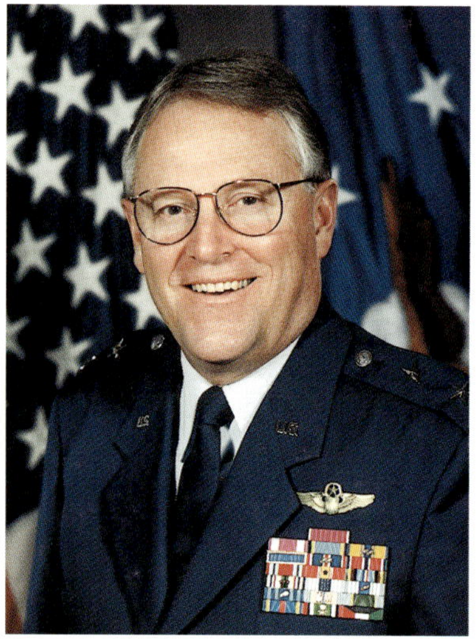

Beginning fiscal year 1999, the Air National Guard possessed 360 F-16 aircraft in 24 separate flying units. That same year the Air Guard deployed 131 different F-16 units with a total of 13,214 personnel on worldwide training missions, exercises and contingencies. Over 820 Air Guard F-16s flew sorties in support of these actions during the year. In celebration of the Air National Guard F-16 force, this photographic volume has finally been produced as a labor of love from the many Air Guard units partecipating in its production.
Mr. Giampaolo Agostinelli presents in this volume his photographs of Air National Guard F-16s from eleven different flying squadrons. I am sure that you will agree that he is one of the foremost aerial photographers in the world. My friend and colleague "Giampy" has not only perfected the art of aviation action photography but knows the difficulties and intricacies of flying fighters. What we say in the Air Guard is that he is a "High-Speed-Low-Drag" flying photographer who brings this exciting flying action to you as if you too were in the cockpit.

Paul A. Weaver, Jr.
Major General, Former Director, Air National Guard

From 1997 to 2003 I had the chance to enjoy very gratifying and unique experiences of flight and of life with several U.S. Air National Guard units and with the "Triple Nickels" of the 31st FW at Aviano.

In 2003 I flew highly interesting, real CAP missions, in the frame of Operation *Noble Eagle* with the last two units I met: 120th Fighter Wing at Great Falls, Montana and 144th Fighter Wing at Fresno, California.

I traveled from one State of the Union to another meeting many people, and now my mind often happens to return to this or that situation, i.e. to all those unrepeatable moments I have lived thanks to those wonderful persons and great professionals whom I met in the Air National Guard.

In the recent years, this Air Force which is also an important complement to the USAF has evolved in the state-of-the-art both technologically and professionally. The many F-16 units are often called for International missions, and therefore employed often on the front line, i.e. in Afghanistan, in the war against the terrorism, in Operations *Northern* and *Southern Watch* and recently, *Iraqi Freedom*.

For this reason they are constantly updated according to the most effective tactics and upgraded with the best technology available.

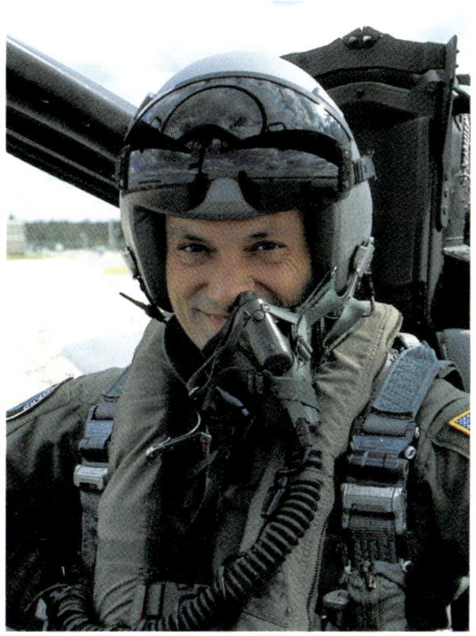

Flying in the F-16s of the units presented in this volume gave me the chance to approach one of the most celebrated and versatile fighters of our time.
This volume is intended to lead the reader in the day-by-day reality of some Fighting Falcon units of this great Air Force, which is the Air National Guard, and of the most important combat unit of the U.S. Air Force in Europe.
I wish to thank Major General Paul J. Weaver, Jr. (former Director, ANG); Lt. Gen. Daniel James III , Director ANG; Maj. Gen. Larry K. Arnold (former Commander, 1st Air Force); Maj. Gen. Craig R. McKinley, Commander, 1st Air Force; John Hoffman; Gary Gault; MSgt. Ronnie Hodnett; Kathleen J. Gereski; Lt. Col. Rod Rodriguez; Maj. Mike Regan; the National Guard Bureau; Maj. Tim Lincoln; Col. Pat Pauli; Col. Phil Skains; TSgt. Heather Pratt; Corine Lombardo; Maj. Donald Arias; SSgt. Scott Farley; Lt. Col. Robert Broderick; Capt. Ed Thomas; Brg. Gen. Mike Worden and Luisa Merlo of 31st FW at Aviano; the Chiefs of Operations; the PAOs; the pilots and technicians of all the units that are mentioned in the various chapters of this volume; and all the other Air National Guard people, certainly not less important, who made this publication possible with their indispensable and precious contribution.
A special mention to my friends Andrea Artoni and Stefano Fogagnolo: without them - I have no doubt - this book would have never existed.
I warmly thank Paolo Bari of the "Nogarazza" Restaurant near Vicenza and to Gabriele Polliero of Due Emme Foto, Padova, and of course to Eric Hehs and Lockheed Martin Tactical Aircraft Systems, Fort Worth, Texas.
Special thanks to Giorgio Pirovano of Grafiche Pirovano. All my photos were taken with Nikon cameras, and the film used was Fuji.

Giampaolo Agostinelli

Facts and figures of the F-16 Fighting Falcon

Since 1981, the F-16 has dominated bombing and gunnery competitions worldwide with a 90 percent win rate. This includes the US Air Force's "Gunsmoke" air-to-ground and "William Tell" air-to-air competitions.

In air-to-air combat F-16s have compiled a 69-to-0 air-to-air exchange ratio with guns and infrared and radar missiles. F-16s achieved the first three combat kills with an AMRAAM missile. In the early 1980s the F-16s effectively destroyed several high-value targets and mobile surface-to-air missile (SAM) sites without a loss.

The F-16 was the workhorse of Operation *Desert Storm*, flying 13,500 sorties with 250 aircraft, including Air National Guard aircraft from the 169th Fighter Wing, South Carolina, and the 174th Fighter Wing, New York. F-16s provided more than 40 percent of US Air Force bomb dropping sorties and delivered 20,000 tons of bombs during the Persian Gulf War.

During the war, the aircraft performed a variety of missions, including Scud chasing and Killer Scout (Fast Forward Air Controller or Fast FAC) and produced the highest mission-capable and sortie rates of any aircraft in theater. The F-16 proved itself capable of operations from austere forward bases and locations to maximize sortie rates and payloads.

Since the Gulf War, F-16 squadrons from several nations have served in peacekeeping operations by enforcing United Nations "no-fly" restrictions over Iraq, Bosnia and Kosovo. In these operations, the Fighting Falcon has served successfully in a number of mission roles including combat air patrol, close air support, enemy defense suppression, reconnaissance and Fast FAC.

The F-16 Fighting Falcon is well over 20 years old; but still an incredibly spectacular powerhouse of weapons, a super ship, without equal. Improvements and upgrades over the years have added weight, but the F-16 is smaller and more nimble than many fighters of its type.

This marvelous war machine with features such as lightweight computers, "fly-by-wire" electronic controls, and an arsenal of high-tech weapons make it a genuine jack-of-all-trades doing so many jobs so well. An airplane in service with many nations, the F-16 can rightly be regarded as one of the finest examples of standard fighter.

The Fighting Falcon is formidable in many situations. In the interceptor role, the F-16 can blast aloft in next to no time and destroy incoming bombers before they can get within letal range.

Over the battlefield, in close air support, the F-16 can destroy enemy tanks and ground positions with a variety of bombs and missiles. For air superiority, the F-16 clears the battle zone of enemy fighters.

In the role of suppression, the Fighting Falcon destroys enemy radars with high-tech missiles allowing other friendly aircraft to fly in safety. With precision strike laser-guided bombs, the F-16 attacks strategic targets such as buildings and power stations.

Giampaolo Agostinelli has flown and driven thousands of miles criss-crossing continental United States documenting the F-16. Thanks to the massive bubble canopy of the F-16 Mr. Agostinelli captures this fast and sleek aircraft on film as only a world-class photographer can do.

You will find "Vipers in The Sky" an interesting and excellent study that is historically rich in information about the phenomenal Fighting Falcon. Flying the F-16 is like nothing you can imagine in your wildest dreams and Giampaolo Agostinelli takes the reader in majestic flight that sometimes is as close as one gets without arming the ejection seat!

John Hoffman

120th Fighter Wing
Montana

120th Fighter Wing, Montana (emblem approved on 6 August 1958). Blue and yellow are the Air Force colors. Blue alludes to the sky, the primary theatre of Air force operations. Yellow refers to the sun and the excellence required of Air Force personnel. The shield's difided background is emblematic of the Wing's mission to be ready at all times to defend the nation against aggressors both day and night. The missile symbolizes the unit's interception capabilities. The snow-capped mountains, rising sun, prairie and river represent the Wing's home state while the rain cloud and sun alludes to the unit's ability to perform in all weather.

186th Fighter Squadron, Montana (emblem approved 27 July 1954). The background represents the blue sky, the bird is a character form of an eagle pacing across and guarding the northern perimeter of the world. The two guns (representative of Montana and the West) symbolize the "shooting down" of any prospective enemy.

186th Fighter Squadron, Montana "Vigilantes".

The roots of the Squadron, and the Montana Air National Guard date back to June 22, 1941 when the German army launched a surprise attack on the Soviet Union. Operation *Barbarossa*, a simultaneous ground and air offensive caught the Russians off guard and nearly decimated their military. By the end of the first week, their Air Force had lost 4,000 aircraft, a number that represented half their force. Joseph Stalin appealed to the United States and Great Britain with a request to be included in the Lend-Lease program. President Roosevelt enacted this program in March of 1941 as a way to provide equipment and supplies to British allies without actually entering the war, which had started in September 1939.

The biggest question became the one of how to get the material to the Soviet Union. Stalin was reluctant to allow American pilots on Soviet soil, but the North Atlantic was controlled by German U-boats. Many Allied ships went down before Stalin agreed to an Arctic flying route.

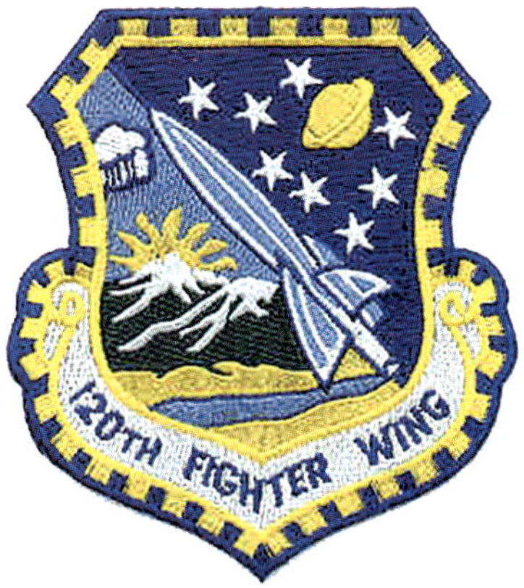

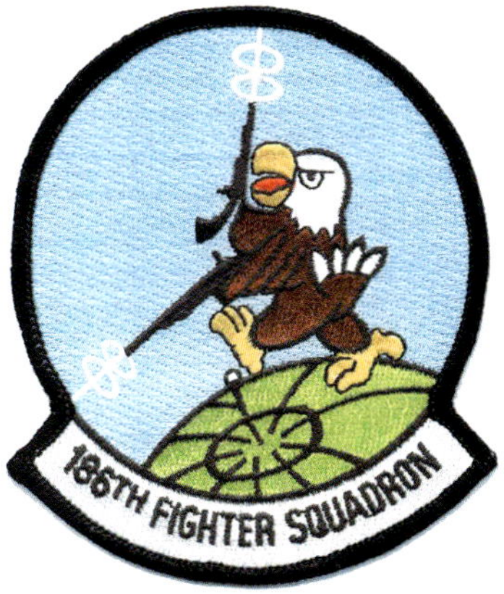

Great Falls, Montana was chosen as the American starting point because of her inland location, over 300 good flying days per year and her alignment with the northern airstrips. On June 22, 1942, the 7th Ferrying Group established headquarters at the Municipal Airport on Gore Hill overlooking Great Falls.

A military airbase was quickly established, named Gore Field after local homesteader James D. Gore. On Sept. 2, 1942, two flights of A-20 light bombers left Great Falls and touched down at Ladd Field in Fairbanks, Alaska. There, Russian pilots picked up the aircraft and flew them home. Known as the Alaska - Siberian route, or ALSIB, this route saw almost 8,000 aircraft provided to assist the war efforts on the Eastern Front from 1942 to 1945.

The 7th Ferrying Group also transported military aircraft to Alaska in support of the 11th Air Force, which was fighting off the Japanese in the northern Pacific theater.

East Base, now known as Malmstrom Air Force Base, was established around the same time. In September of 1945, the military installation at Gore Field, by then known as the 557th Army Air Forces Base Unit, was closed down. Following the war, each state was authorized one air unit. Lt. Col. Willard Sperry, a decorated combat pilot stationed at East Base, began building the Montana Air National Guard at Gore Field.

On June 27, 1947, the 186th Fighter Squadron was activated and federally recognized. Within two weeks, six P-51D Mustangs arrived. On April 1, 1951, the unit was activated for Korea. Personnel were sent to Moody Air Force Base, Georgia, and 10 F-51s were shipped to Korea.

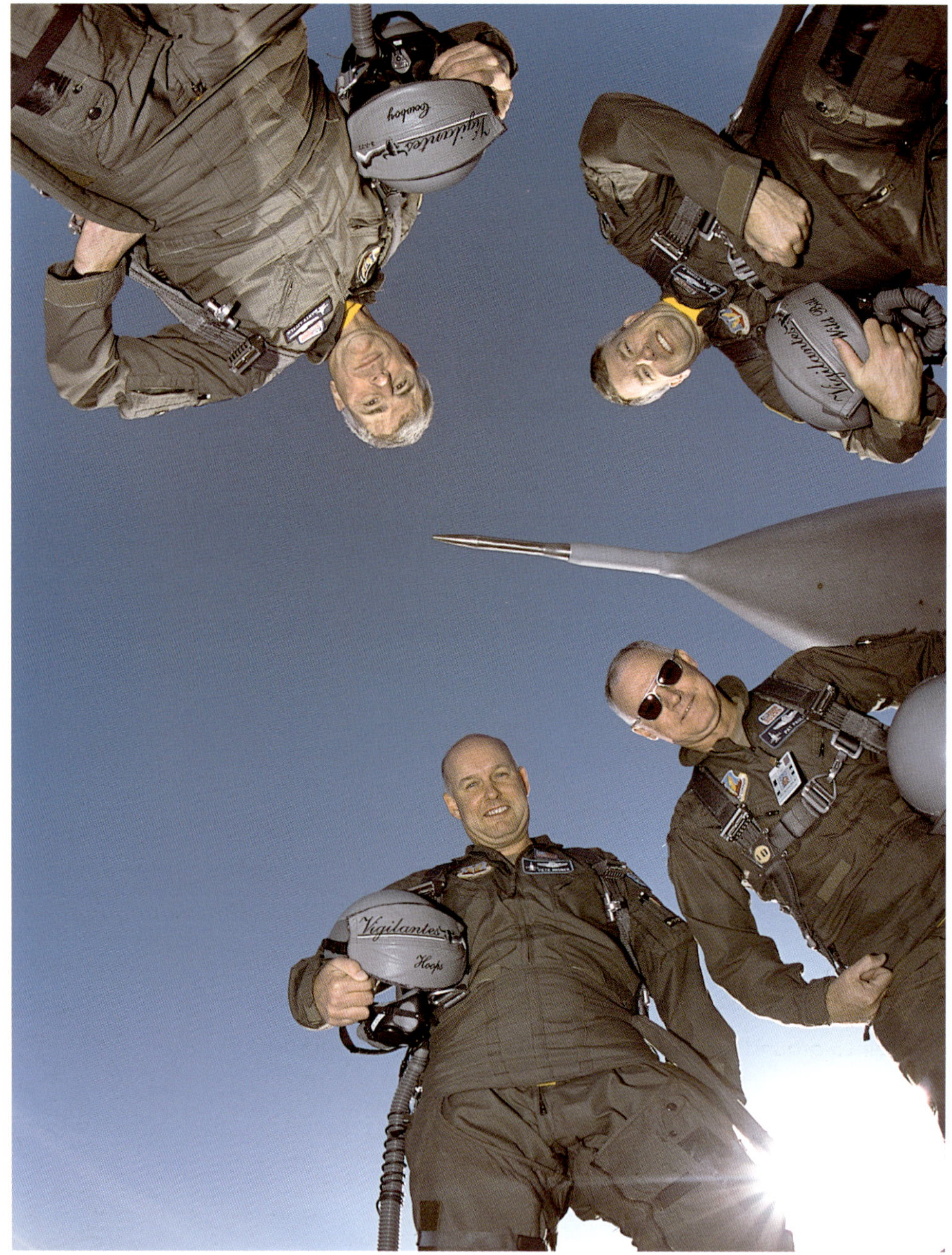

On the right
120th FW pilots, including CO Col. Pat Pauli (second, counter-clockwise), pose in front of a Viper nose cone.

The unit received its first jet aircraft in 1952, a T-33 Shooting Star. The first jet fighter landed at Gore Field a year later in the form of the F-86A Sabre Jet, the first F-86 assigned to an Air National Guard squadron. One is currently mounted as a weather vane overlooking Interstate 15 at the airport exit.

The squadron was redesignated the 186th Fighter-Interceptor Squadron and adopted the "Charlie Chicken" patch. The F-89 Scorpion arrived in 1955 and the Montana Air National Guard was redesignated the 120th Fighter Group (Air Defense).

An F-89J is mounted above the front gate of the air base, that particular aircraft has the distinction of being the only F-89 to have ever fired a Genie nuclear missile. October 1, 1958 marked the beginning of the unit's commitment to five-minute runway alert, a task that would last for 38 years. The arrival of the F-102 Delta Dagger in 1966 ushered in the supersonic age. In 1972, the unit was redesignated the 120th Fighter Interceptor Group and assigned the F-106 Delta Dart, the first Air National Guard unit to receive this aircraft. With the F-106, the 120th competed in and won its first William Tell, a live-fire missile competition held at Tyndall Air Force Base, Florida. The F-16A/B arrived in 1987 and the unit became the 120th Fighter Wing in 1995. The 120th completed its conversion to the F-16C/D and has replaced the air defense mission with one of general purpose/air-to-ground as part of the Expeditionary Aerospace Force.

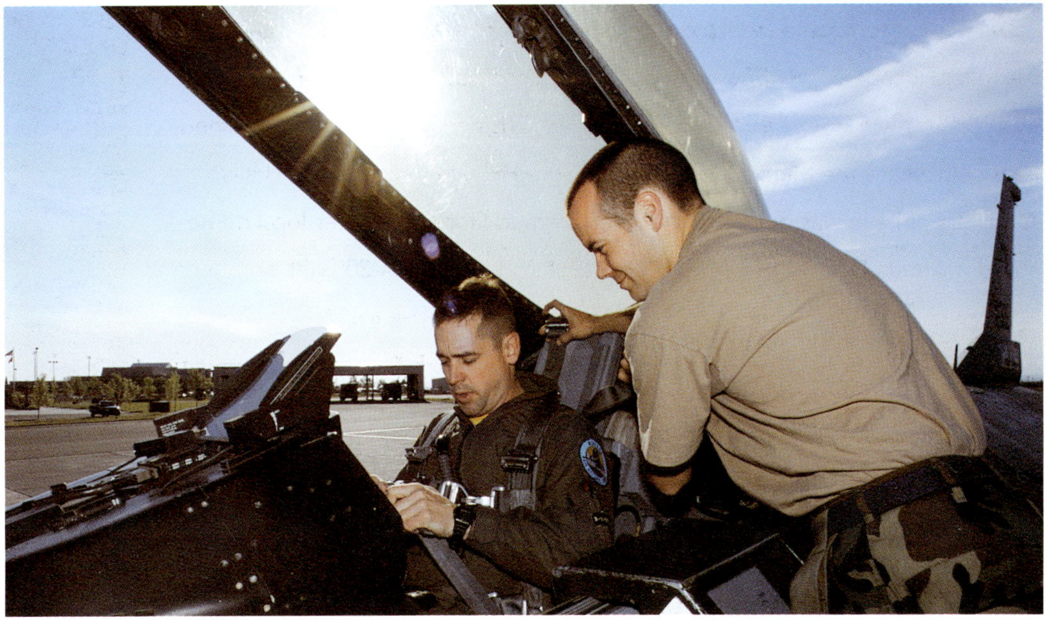

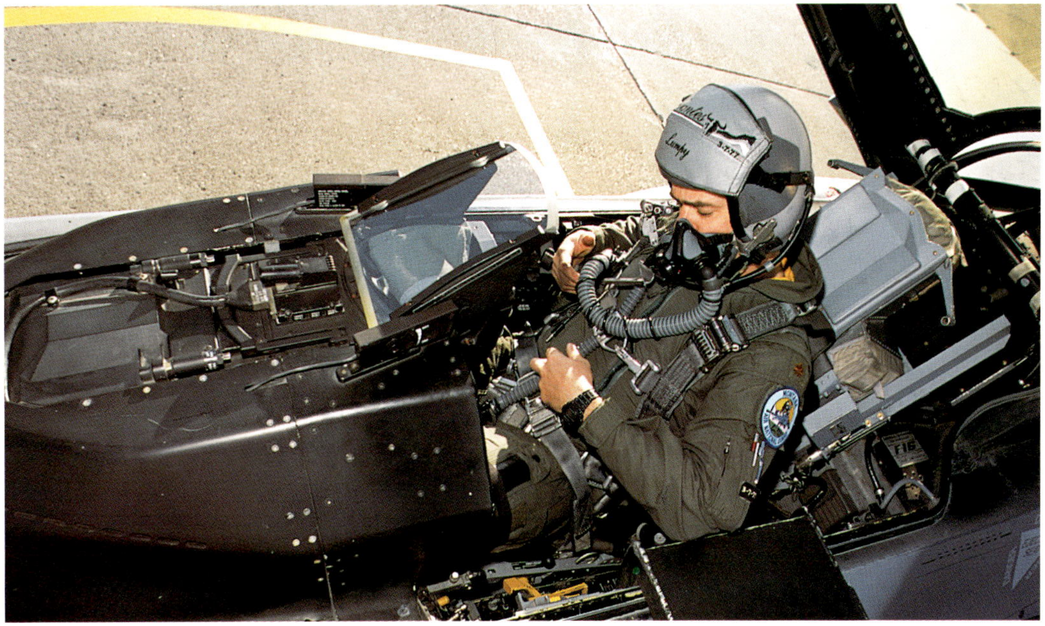

On the right and above
Two 186th FS pilots perform their pre-fligh checks before a training mission.

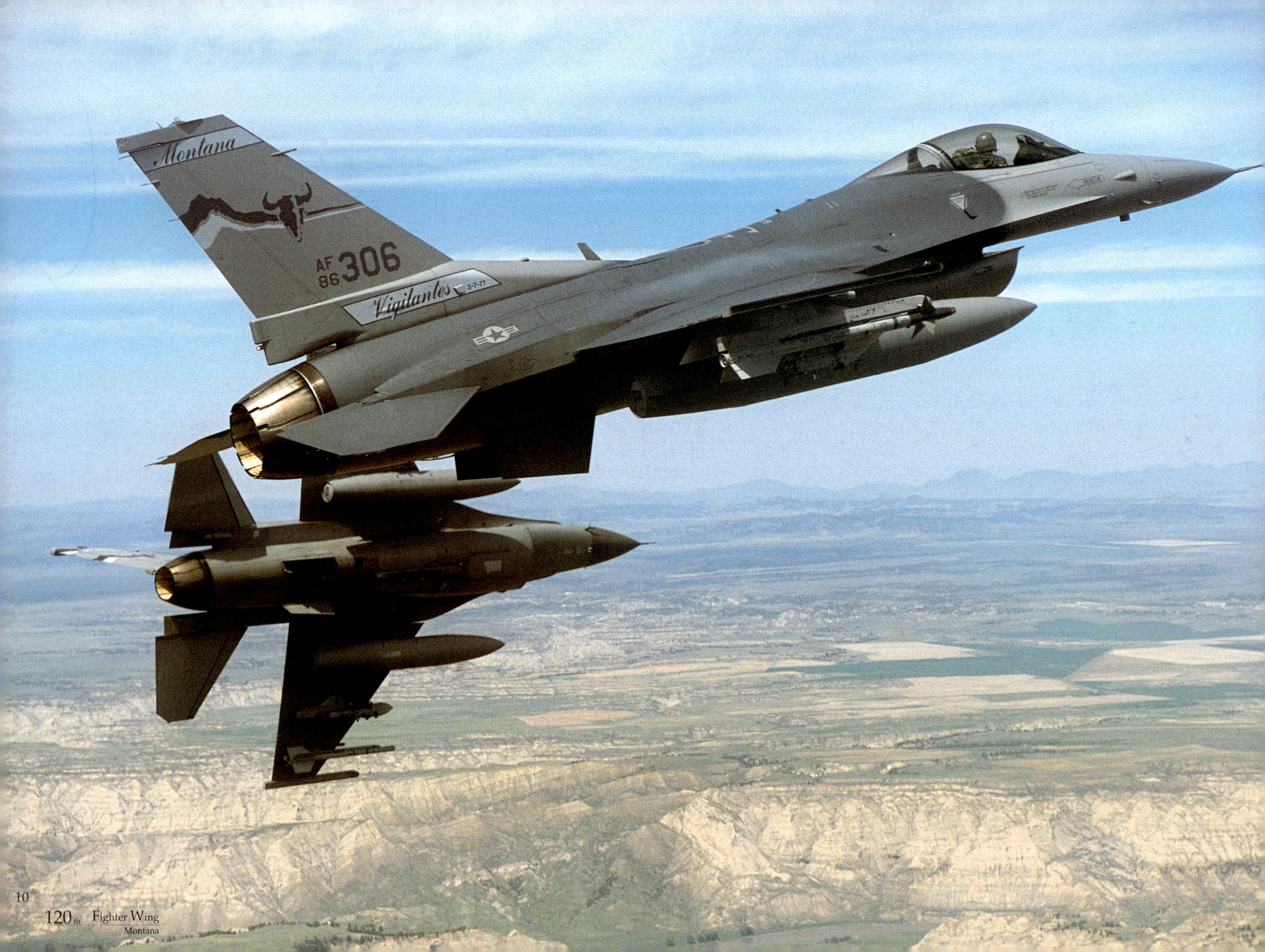
Montana
AF 86 306
Vigilantes

A NIGHT CAP IN THE IRAQI SOUTH NO-FLY ZONE

BY CALL SIGN "MULE"

The 120th Fighter Wing was well prepared for the deployment to Prince Sultan Air Force Base, Saudi Arabia to enforce the Southern No-fly Zone over Iraq, and eagerly waited getting under way. For the first few combat sorties there was some nervous anticipation in the air. The first week of air defense flying was accomplished during the day and the second week, nights and then back to days.

The Maintenance personnel had the same anticipation and it could be felt when doing the walk around at the jet. Their support made you feel like even though they were not physically in the plane with you they understood and waited eagerly for your return. Most of the time there were five or more shaking your hand at the jet and wishing you a safe flight. The first night sortie for my wingman and I still carried that same anticipation for the both of us. There have been many sorties over the years that pilots have encountered much worse but as our first time it will never be forgotten.

Just like any other flight it all started with mission planning and then the mass brief. Thereafter we split up for our own squadron brief and then a two-ship brief. Hitch (my wingman) and I were finally ready to go. We stepped to Life Support to suit up and check out our night vision goggles and our personal weapons. Life Support gave us a ride to the jets that were parked on the other side of the airfield and provided some small talk along the way.

After arriving at the jet and doing the walk around the Crew Chief asked me if I would fly a U.S. flag over Iraq for him and some others. It seemed it was the least I could do, since he had been busting his tail in the hot sun all day to ensure I had a good jet. After starting the engine and going through the standard checks we were ready to taxi. While arming at the end of the runway my electronic counter measures (ECM) pod was having problems but seemed to work fine after cycling the power. Hitch was now through arming and was ready to go which was evident by him going to flash on his external lights. We took the runway and went all lights out for takeoff. After checking the engine I lit the afterburner and away I went with my wingman in trail.

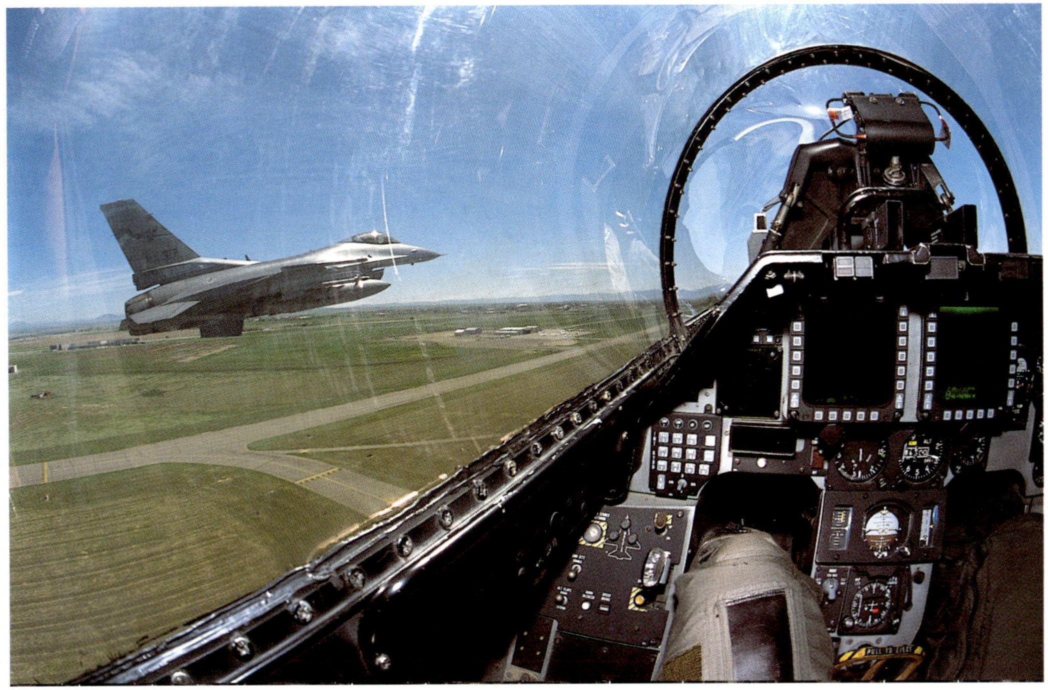

Above
Formation take-off from Great Falls Airport.

In the page on the left
A dynamic split of a two-ship formation.

At the end of the runway we proceeded to a vertical climb, looking behind us for any surface to air missiles coming our way. After we were in the clear we donned our Night Vision Goggles (NVG's) and Hitch joined to the briefed line-abreast formation. During the flight north to Iraq I was amazed by all the cars I could see on the highways with the NVG's.

It was a little hazy from all the dust but once above 20,000 feet it seemed to dissipate. Just before entering Iraq we needed to re-fuel from one of the many tankers that took off before us. This almost seemed a bigger threat to us at the time because of the amount of aircraft in the small piece of sky at the same time. The goggles were worth their weight in gold. After taking on a full load of fuel, Hitch and I headed in to Iraq to replace two of our squadron buddies. Assuring we were lights out we located them using our situational awareness data link (SADL) and passed them in the night as they returned to base.

Talking with them on the secure radio they informed us it was fairly calm with some antiaircraft artillery (AAA) once in a while in the distance. Our mission was to protect defenseless, but high value assets operating in the Southern No-fly Zone and we were to do a combat air patrol (CAP) over the eastern side.

On the way north there were miles of oil wells burning that gained down our goggles when looking directly at them. After a few minutes droning north we had to start picking our way through several suspected surface-to-air missile sites and cities.

I could tell Hitch felt the same nervous adrenaline as I did from his tone of voice on the radio. My head was constantly moving and looking down for any type of surface-to-air weapon being fired at us. At this point I remember wishing my goggles had a larger field of view.

We began to work farther north into Iraq when we both called out some surface-to-air missiles left eleven o'clock. Even though it is hard to judge distance with the goggles we knew they were out of range from us based on where they came off the ground. The radios came alive and you could hear another pilot defend-ing and making a call of his position. This brought a strange sense of realism to my cockpit and I was glad he was able to de-feat the missiles. As we preceded north my RWR started to chirp differently than the usual nuisance spike and my wingman also called, "Mud spike twelve o'clock." I elected to turn west and looked over my right shoulder for any launch.

When fairly assured nothing had been launched we turned south where our spike went away. We needed to get north to help protect other assets so we headed back in that direction where we encountered that same threat, and turned south again. At this time we were getting pressed for time to get in position so we tried a third time. In the same location we re-ceived the same indications and turned to the west. I then turned back to the east to keep the spike off my left wing so I could see anything coming off the ground. Just as I rolled out eastbound I received a missile launch on my radar warning receiver (RWR).

I deployed some chaff and started to maneuver my jet. In a des-perate search for a visual confirmation Hitch called defending east. Our SADL had just quit working so we relied on the radio for deconfliction from one another.

In the page on the right
In close formation over the Glacier National Park.

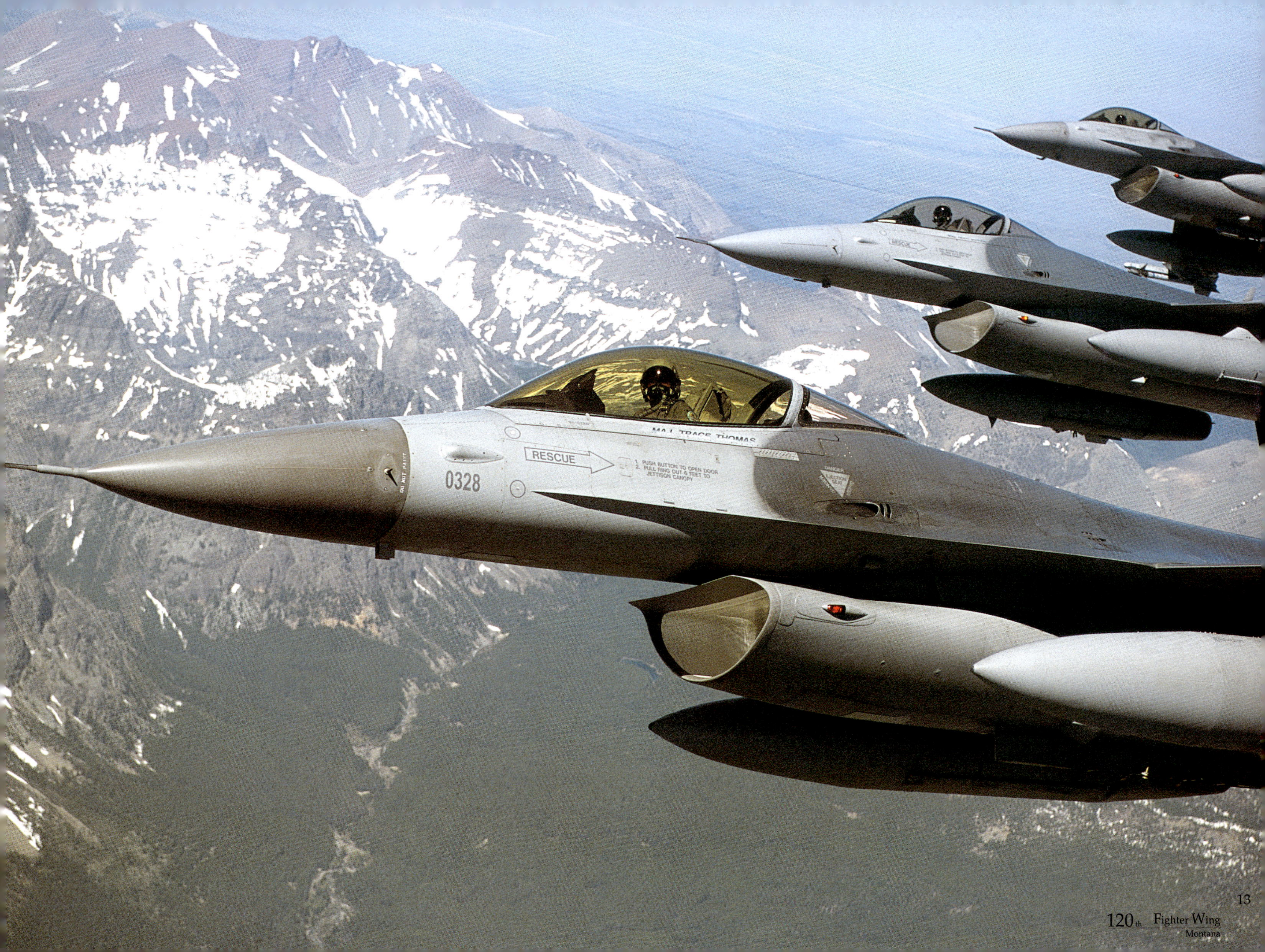
RESCUE
0328

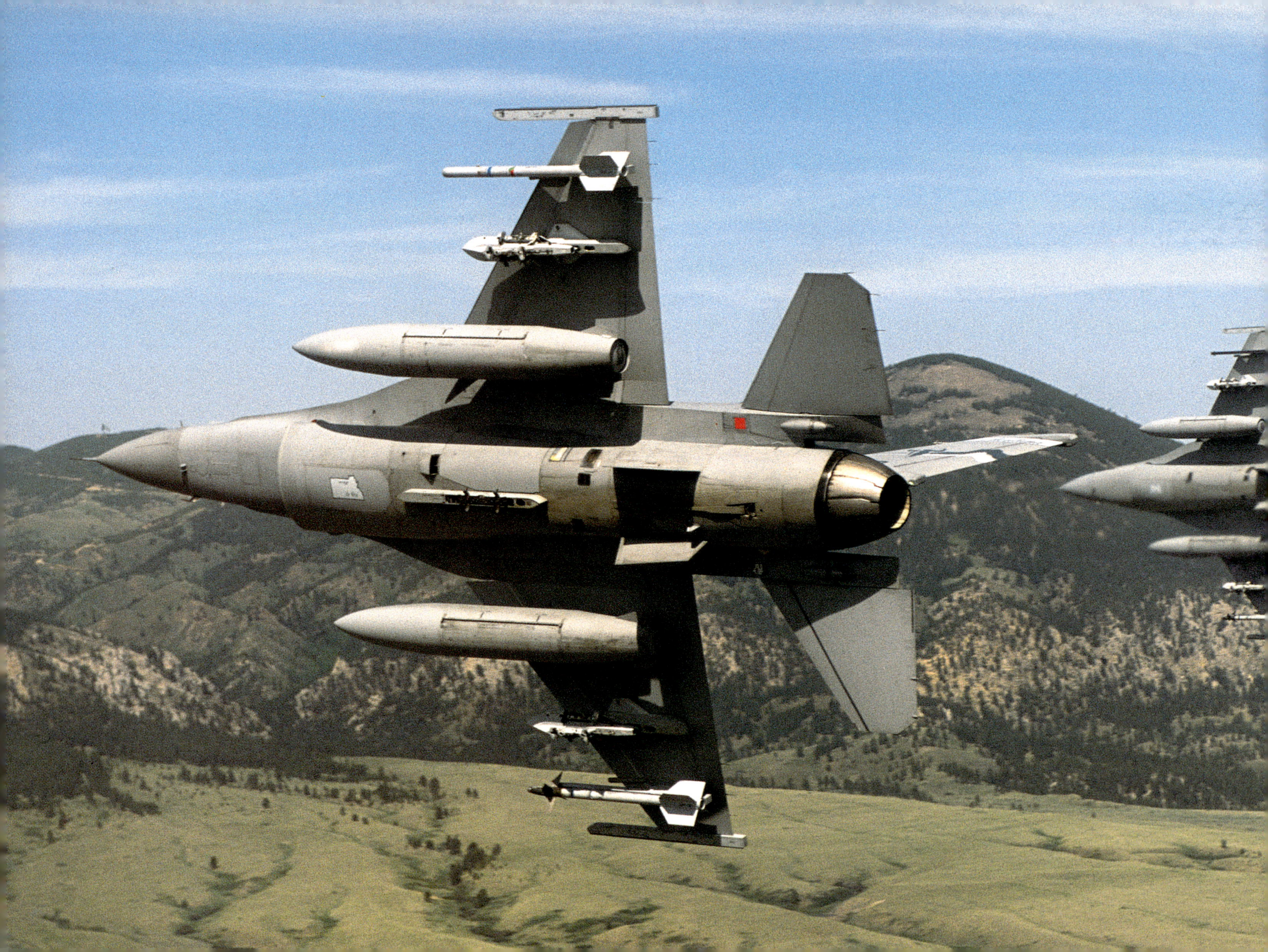

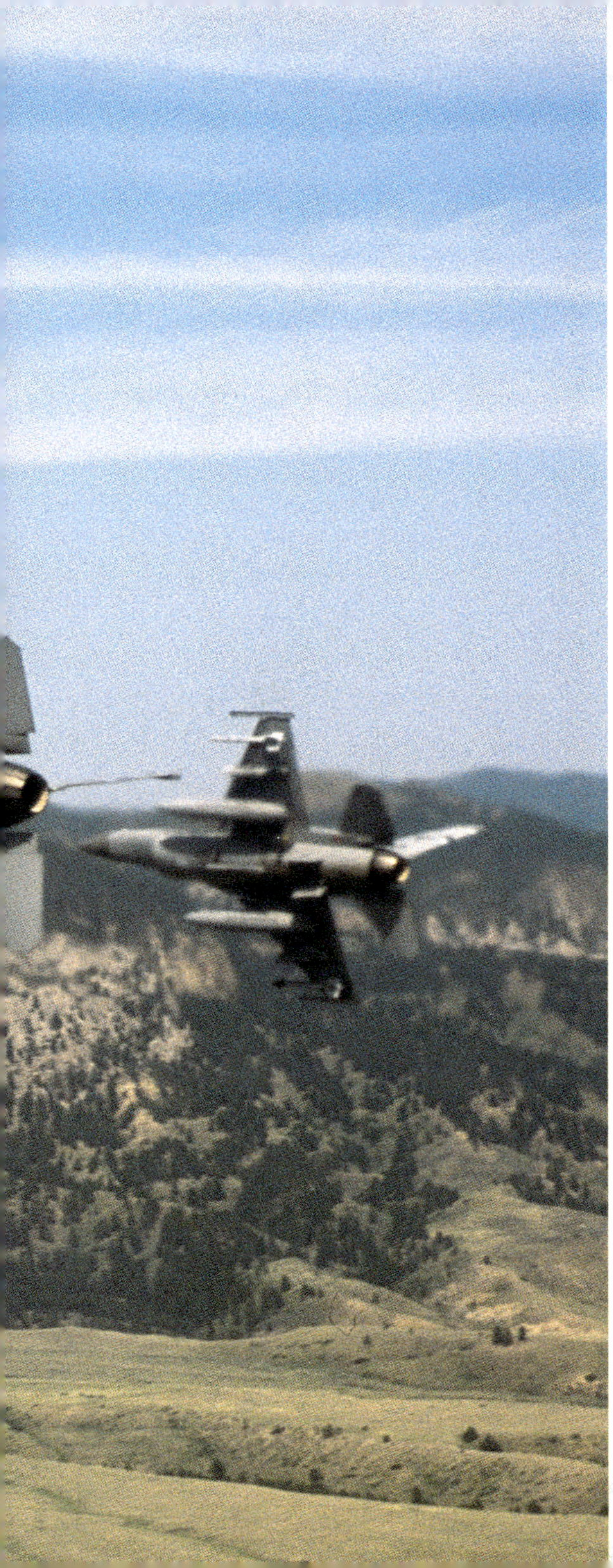

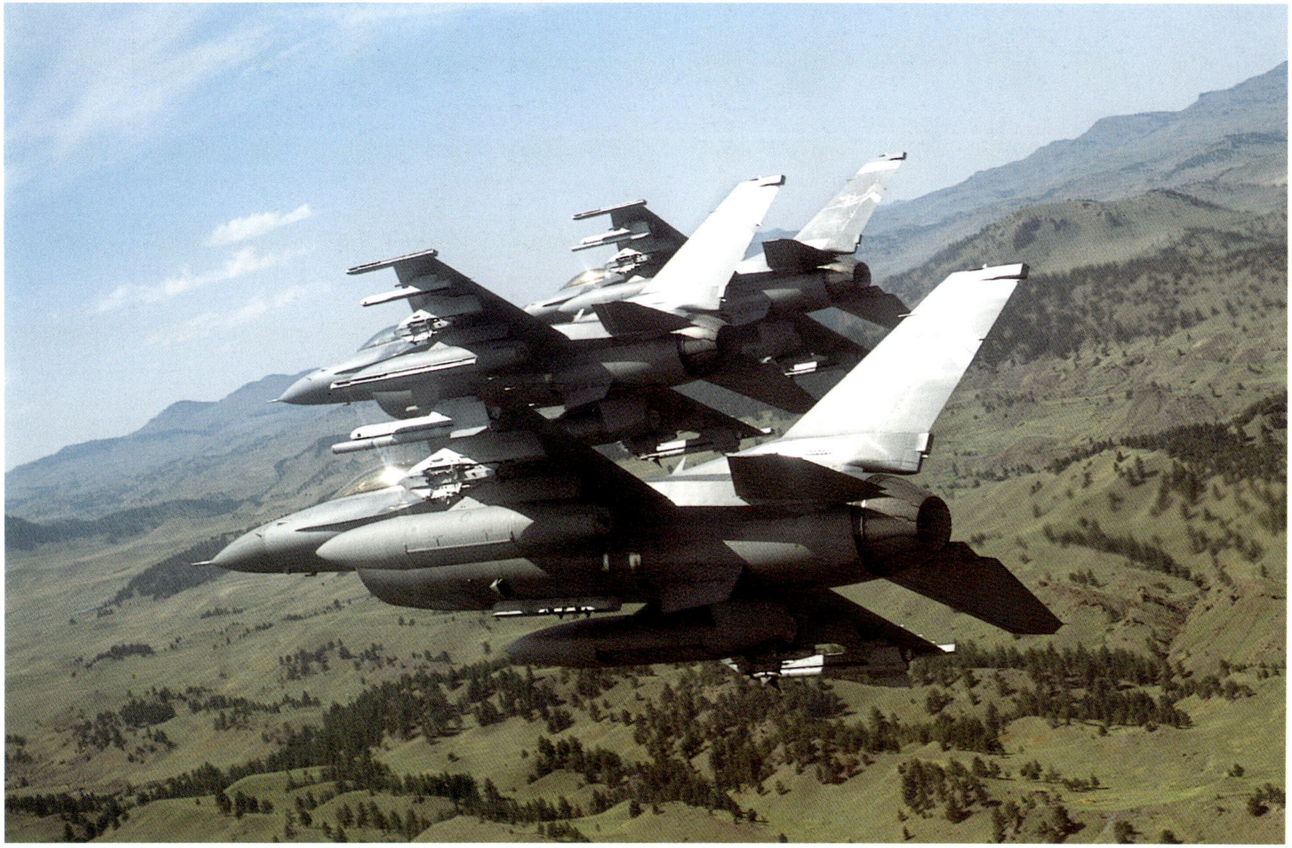

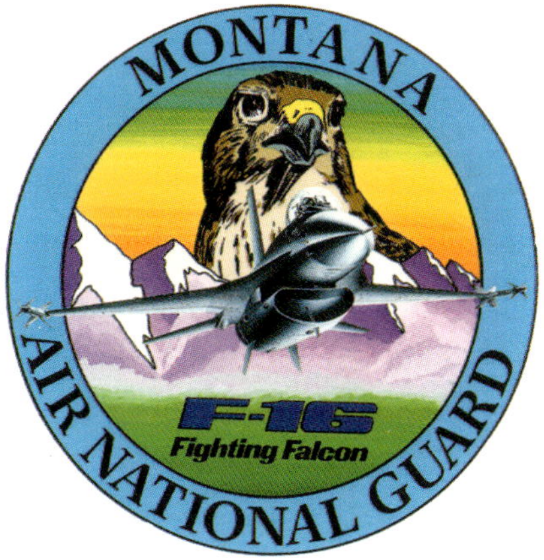

Above and on the left
A spectacurar echelon turn over the rolling hills as it is seen from the far left wingman station.

On the left
A non-official shoulder patch worn by 120th FW pilots shows the F-16, the falcon and the Montana landscape.

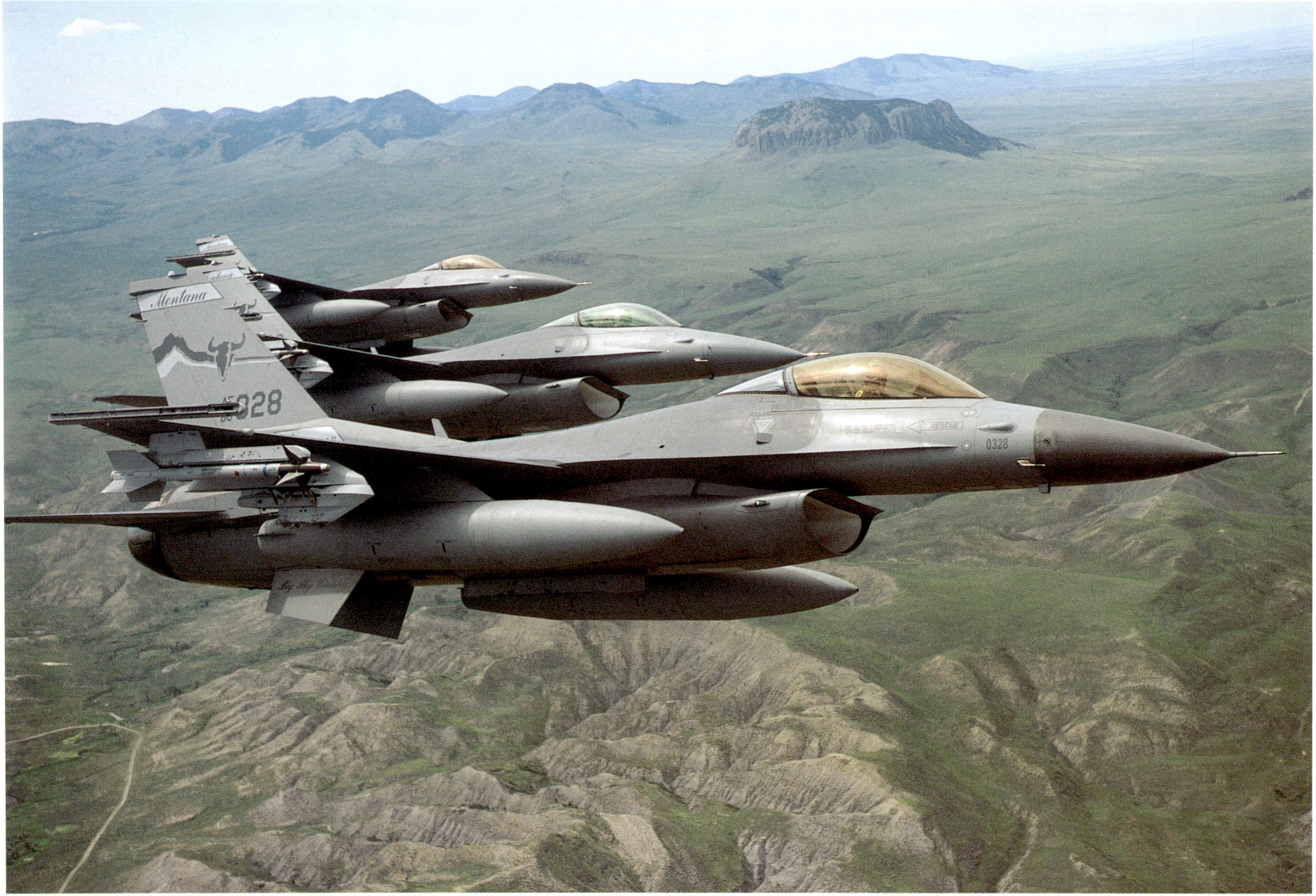
Montana
328
0328
RESCUE

It was nearly impossible to visually keep track of each other while trying to find the missiles leaving the ground. Just as I began to maneuver I saw two missiles coming off the ground one after the other. The missiles produced a bright orange glow and burned out at what I figured to be twenty thousand feet. Just after missile burnout I lost the spike but I continued to maneuver and kept the last known position of the missiles off the left wing.

It felt like forever just wondering where the missiles were going when another flight called out explosions well above us. That call gave me some relief knowing that the missiles had missed. Now the next task was to find my wingman.

After a couple of radio calls we were able to locate each other on radar and join back up. He ended up well north of me while defending against that same threat location. After clearing the area we dressed up our flight and continued the mission. For the rest of the night the surface to air threat seemed to shut down. It was almost as if someone just turned of the switch and all the shooting stopped.

At the end of the mission we headed south out of Iraq. We topped off our fuel on the way back to the air base and the radios fell almost silent for the rest of the night. After landing Hitch and I finished our two-ship debrief and the mass debrief where the adrenaline still continued to flow. During that sortie I saw at least eight surface-to-air missiles and uncountable AAA at all altitudes.

Even though I feel we were very prepared for the mission in Iraq it still opened my eyes. The people that continue the mission day after day and are there when called are real professionals and deserve to be commended.

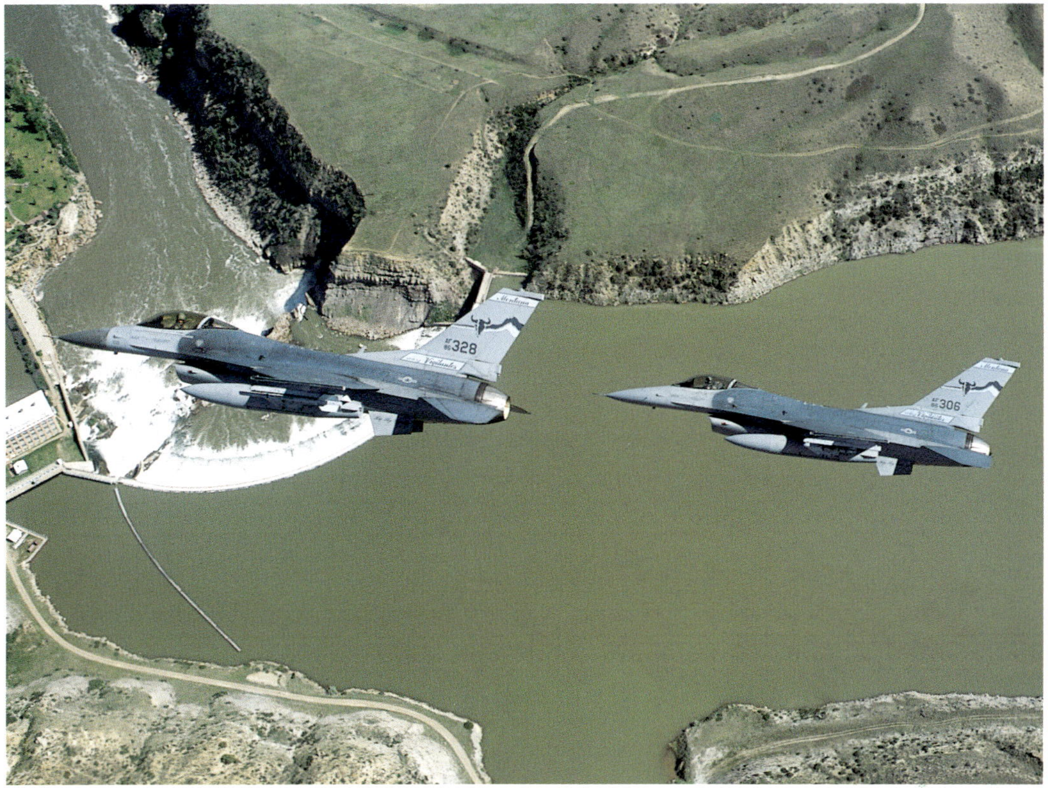

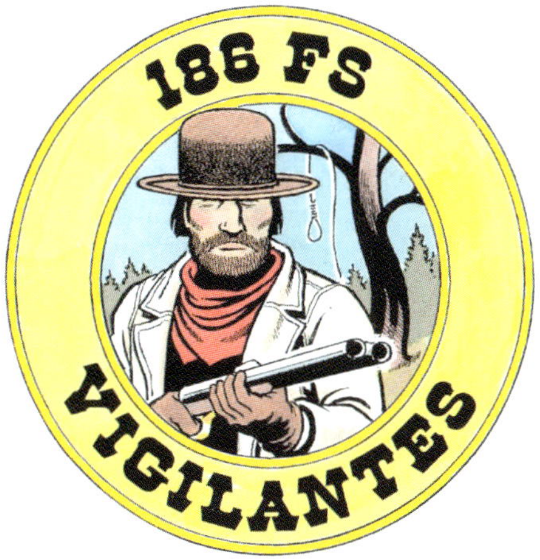

Above
Two Vipers overfly the Great Falls of the Missouri River.

On the left
Another unofficial patch worn by 186th FW pilots.

In the page on the left
The pure lines of the F-16 show off in this neat formation.

In the page overleaf
Turning over the deep green of the Big Sky Country.

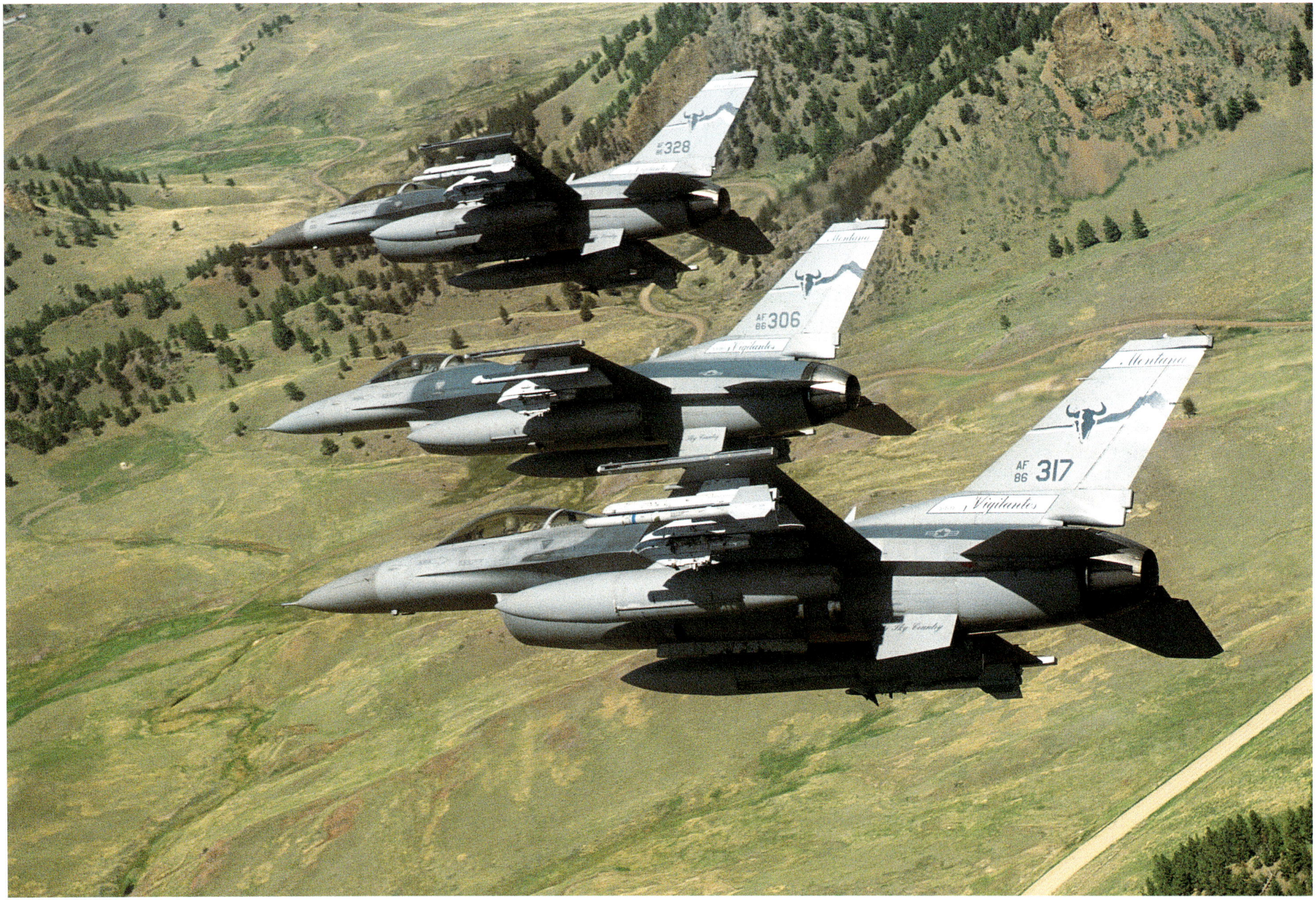
AF 86 328
AF 86 306
Vigilantes
Montana
AF 86 317
Vigilantes
Montana

144th Fighter Wing
California

144th Fighter Wing, California (emblem approved 10 February 1954). Blue represents the sky area of combat. The three stars represent the states: Nevada, Utah, and California. The horizontal flash of lightning symbolizes air-to-air destructive power and the diagonal flash, air to ground destructive power.

194th Fighter Squadron, California (emblem approved 6 Novembre 1951). The griffin is a mythical creature, one-half lion and one-half eagle. It is renown for its aggressiveness in battle and for the keenness and courage displayed in its defense of cave or castle. The griffin, emitting flames, and with flashing talons and beak, screaming defiance at those who would encroach upon its domain, is emblematic of the like qualities of the 194th Fighter Squadron.

194th Fighter Squadron, California "Griffins".
The origin of the 144th Fighter Wing dates back to 4 April, 1948, barely six months after the formation of the Air National Guard in September of 1947. On that date, the 61st Fighter Wing received federal recognition at Alameda, California, followed by activation of the 194th Fighter Squadron on 25 June, 1948. The 61st Fighter Wing was redesignated as the 144th Fighter Bomber Wing on 1 November, 1950. On 7 July, 1955, the Wing was redesignated as the 144th Fighter Interceptor Wing, with the 194th becoming the 194th Fighter Interceptor Squadron. Effective 16 March, 1992, the 144th Fighter Interceptor Wing was redesignated as the 144th Fighter Wing (144th FW), with all related Fighter Interceptor Groups and Squadrons becoming Fighter Groups and Fighter Squadrons. In 1979, the 144th Fighter Interceptor Wing was reassigned from the Air Defense Command to the Tactical Air Command, and in 1992, 144 FW was reassigned to Air Combat Command.

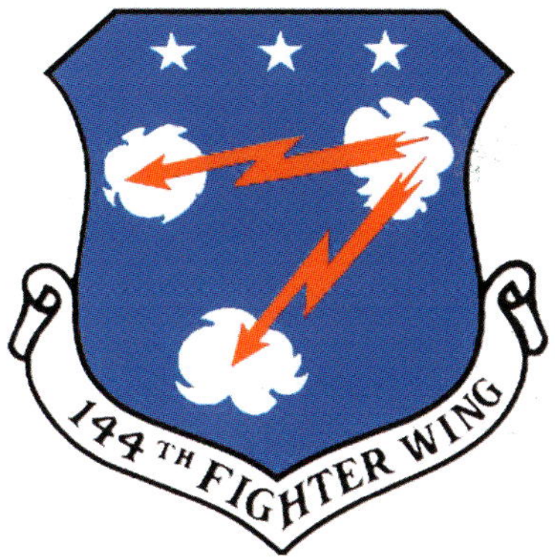

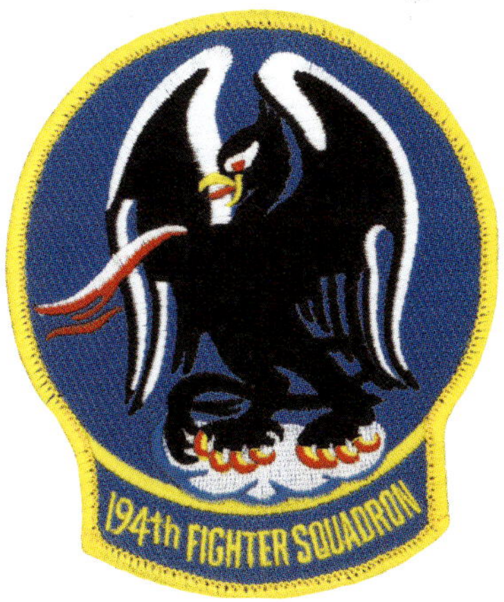

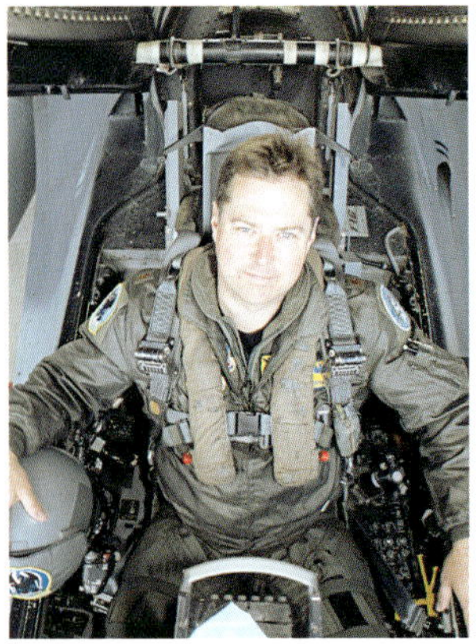

On the left
Maj. Mark Crump sitting in an F-16C, ready for a "scramble" take-off.

On the right
Armed fighters are always guarded by armed ground crew.

The first aircraft assigned as of June 1948 included two T-6s, one B-26, one C-47, one L-17, and two P-51s. The P-51D and later the P-51H were flown until 31 October 1954. During these early years, the unit earned prominence as one of the Air Force's most respected aerial gunnery competitors. In June 1953, while still flying the P-51, the unit qualified for the first all-jet, worldwide gunnery meet. Using borrowed F-86A Sabre jets, the 144th, which represented the Air National Guard, placed fifth in competition. On 1 November, 1954, the 194th accomplished the transition to the F-86A and at the same time it relocated to Fresno, followed by the wing in 1957. The site of the Fresno Air National Guard Base has been used for military aviation since World War II, when it was known as Hammer Field. On 1 April ,1958, the transition was made to the F-86L. On 1 July, 1964, the unit began flying the F-102A, and on 25 July, 1974 the 144th brought the F-106 into service. On 1 January, 1984 the F-4D became the wing's new aircraft, and on 1 October, 1989 the wing completed the transition to the air defense version of the F-16A Fighting Falcon. In September 1995 the transition to the F-16C became complete. The wing's current inventory includes 18 F-16Cs, one two-seat F-16D and one C-26A transport.

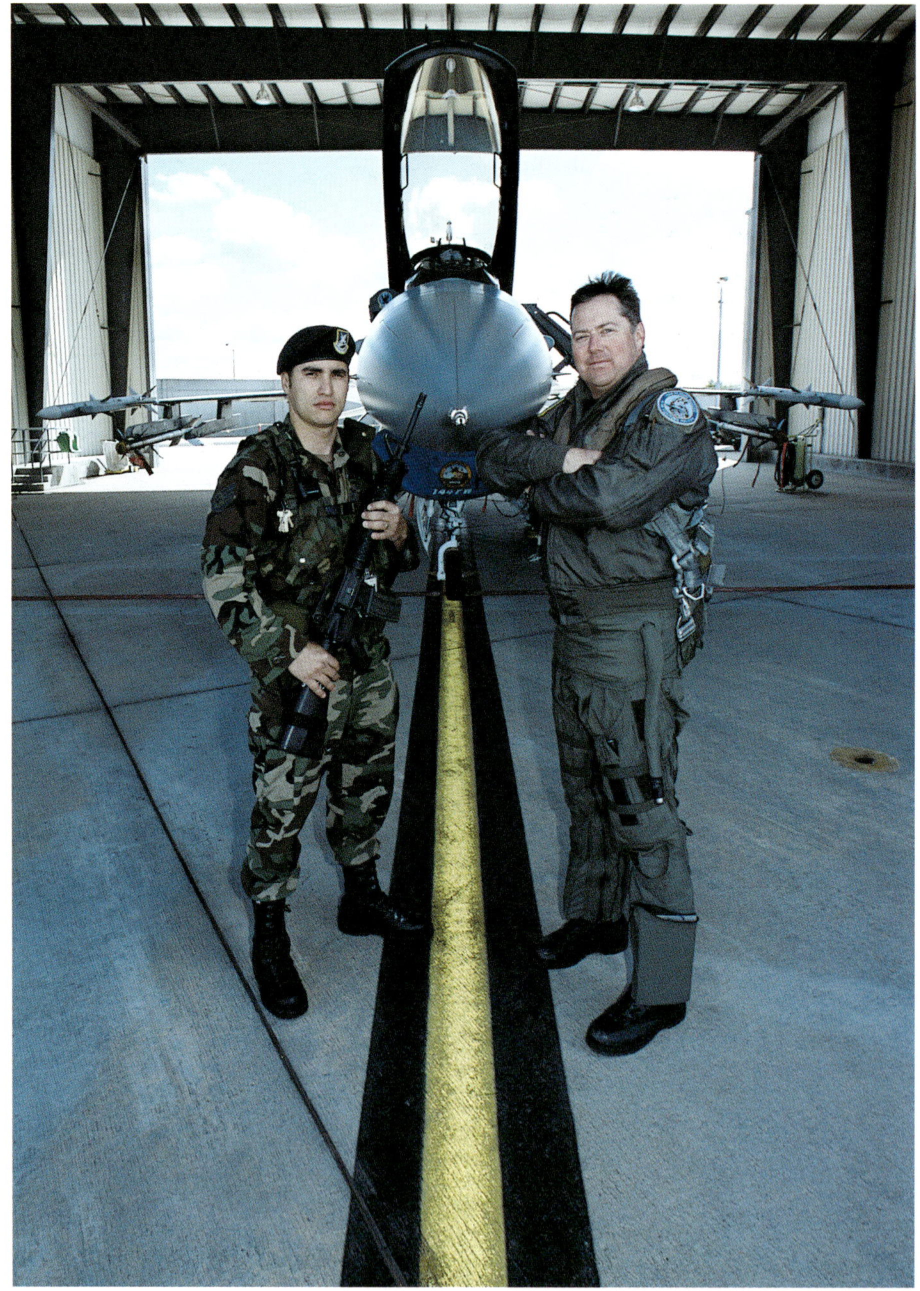

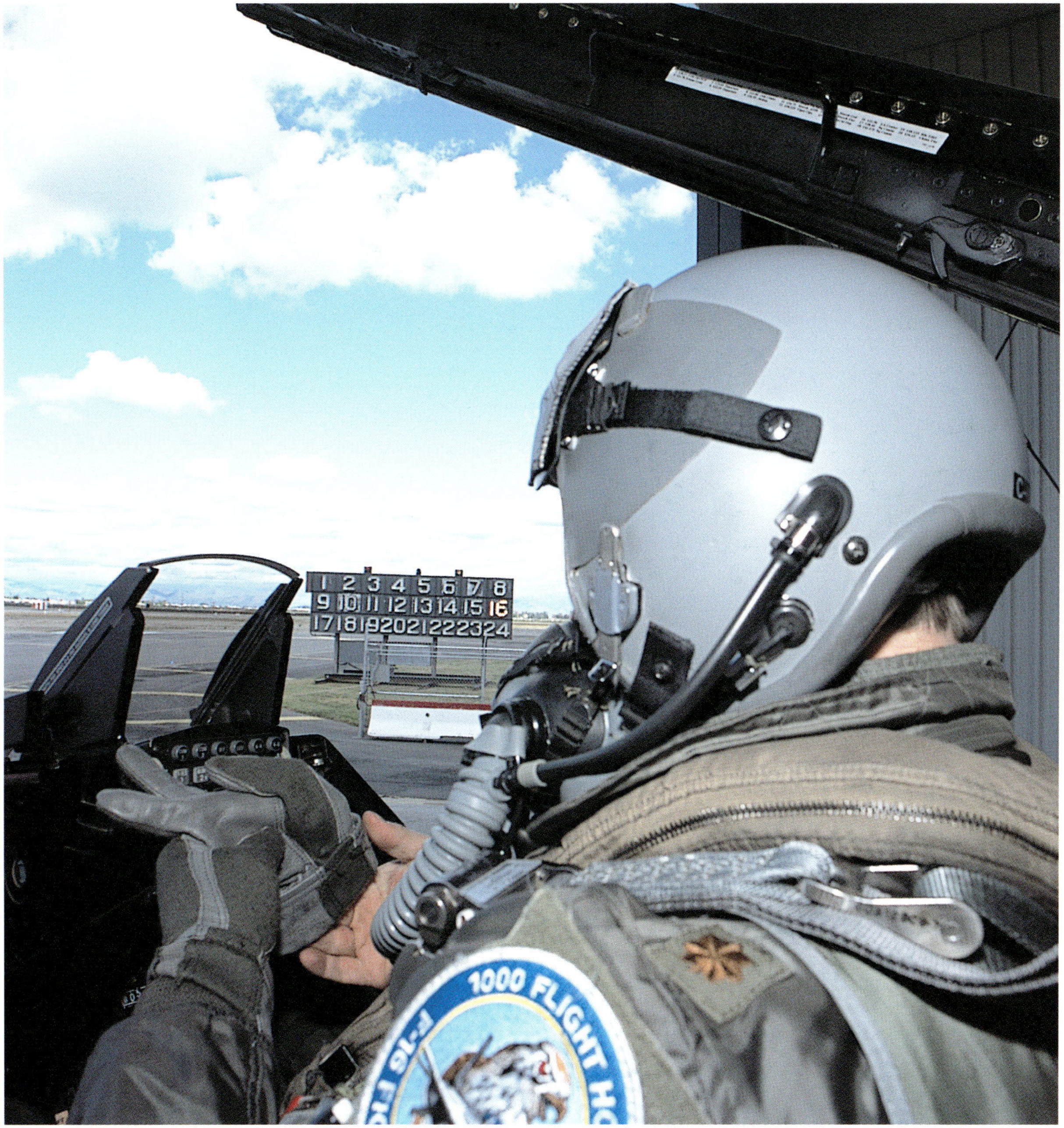

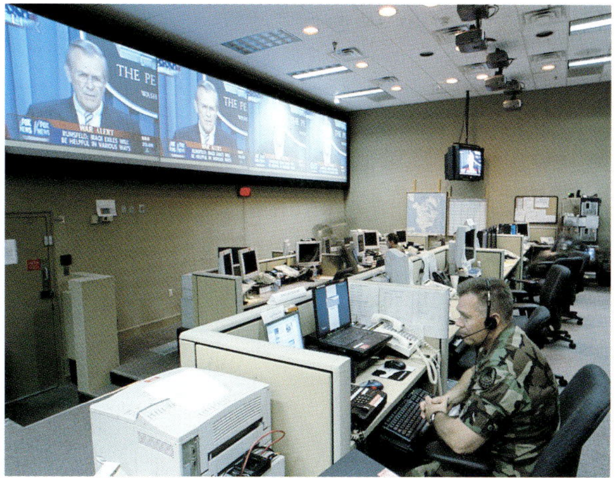

On the left
Under the shade of an open shelter, a 144th Figher Wing pilot is ready for engine start and "scramble" take-off in a mere three minutes.

Above
The OPS Control Room of Southeast Air Defence Sector, Tyndall AFB. The picture was taken during Operation *Iraqi Freedom*.

Below
The 144th Flight Surgeon is greeted after his last flight in the Viper.

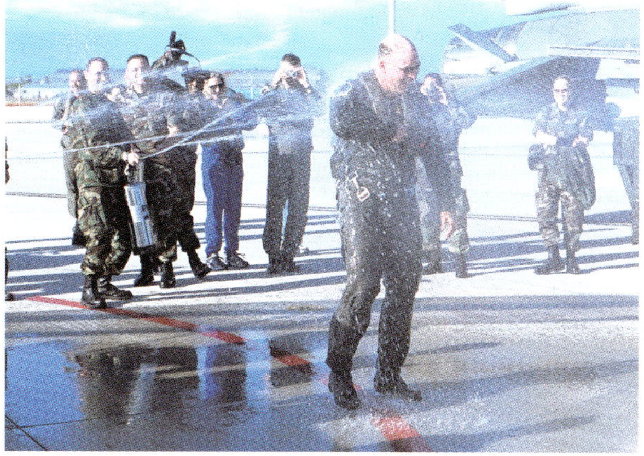

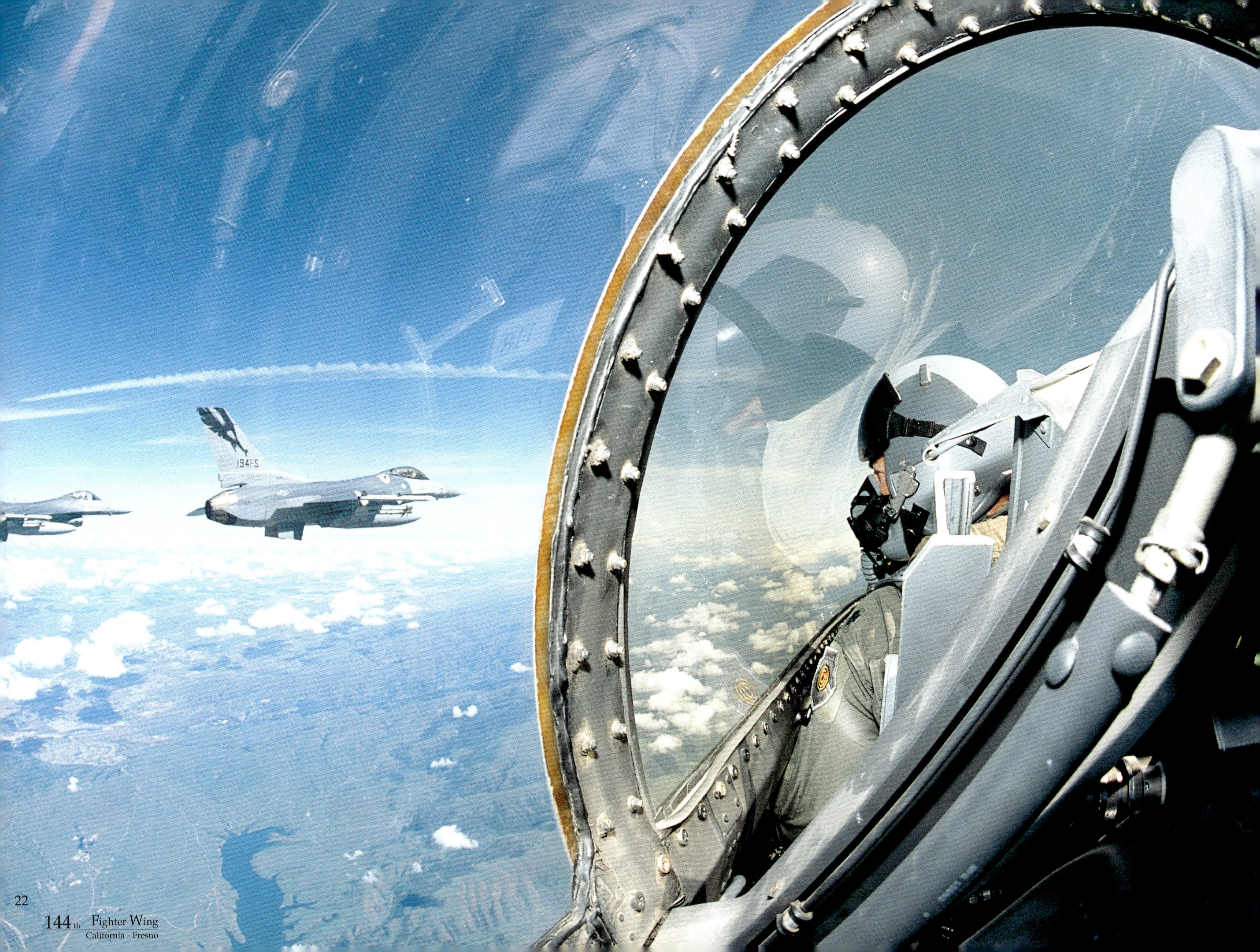
194FS

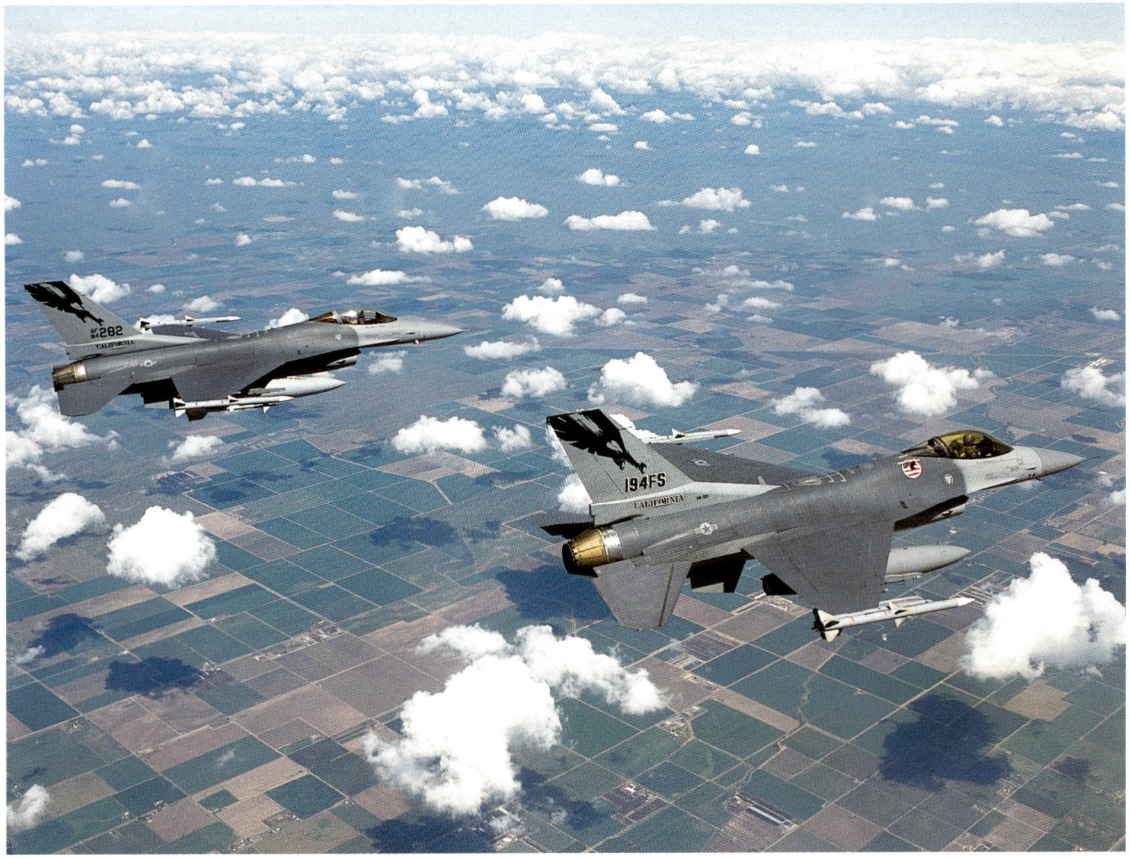

Above
AIM-120 AMRAAM active radar-guided missiles and AIM-9L IR-guided missiles for self-defence are the stardard weapons for a CAP.

In the page on the left
Mid-air rendez-vous with the San Francisco CAP returning from their mission.

THE TIP OF THE SPEAR

BY LT. COL. RYAN ORIAN AND MAJ. ROB SWERTFAGER

The 194th Fighter Squadron is the home of the "Griffins", America's most experienced air-to-air F-16C fighter pilots.

Whether it is Homeland defense or deployed overseas on an Air Expeditionary Force, the "Griffins" of the Fresno guard portray excellence on all levels.

Currently 32-pilots strong, with an additional 12 life support troops, the squadron has witnessed increased operations tempo never seen in the unit's history. Not only has the alert mission increased 100 percent since September 11, 2001; we have also participated in Air Force One escort duty, World Series and Super Bowl Combat Air Patrols.

One would think that with such an increased work load the mission accomplishment rate might slip. Not so, rather, the squadron's performance has increased.

With a significant amount of the squadron's training flights now flown for homeland defense, we have been challenged to meet our own combat training numbers.

The squadron has done an incredible job with its flexibility and availability to accomplish both NORAD's air defense mission and Air Combat Command's training requirements. The dedication of the "Griffins" has kept America's skies safe while enabling us to continue to contribute in the AEF cycle. Of course none of this would be possible without the continuous support that the squadron receives from the 144th Fighter Wing. Our success as a fighter squadron can be directly traced to the folks who help us on a daily basis.

A prime example of the 194th Fighter Squadron's ability came in the form of a no-notice NORAD Alert Force Evaluation at our detachment in Riverside. During the scramble, our pilots expertly intercepted an unknown aircraft and prosecuted NORAD's assigned mission flawlessly. Their effort resulted in an "Outstanding" rating in aircraft employment, the highest rating possible.

Here are a few little known facts about the "Griffins" of the 194th Fighter Squadron.

The average F-16C flight time for our pilots is 1,300 hours, whereas the active duty squadron has 500-700 hours. Over 88 percent of our pilots have combat experience over Iraq, including four pilots who flew during Operation *Desert Storm*. 16 of our squadron pilots are Traditional and they are averaging a 200-hour work month while not being mobilized. Our Life Support office is phenomenal, providing the flyers with state-of-the-art equipment and expertise.

We have equipment other squadrons only dream about. Without them on the team we would not be able to operate in the demanding flying arena.

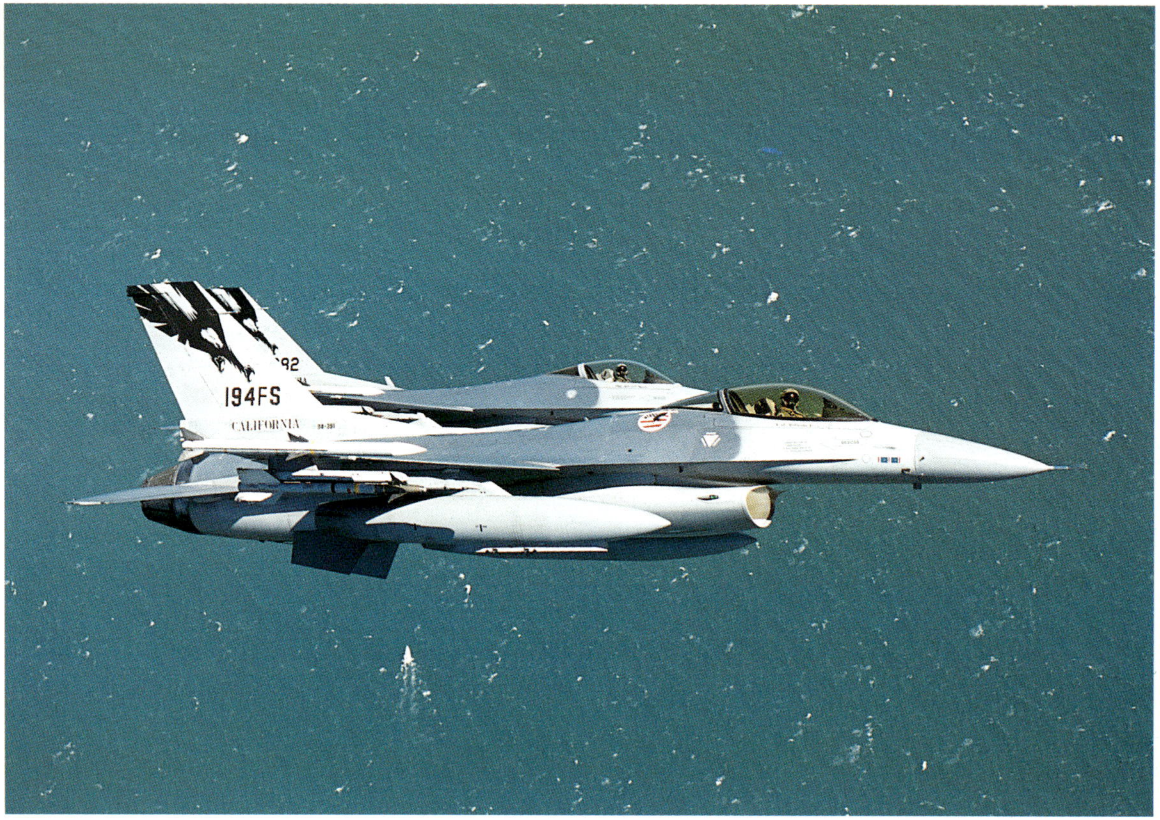

Above
Two 144th Fighter Wing F-16Cs in close formation over the Pacific Ocean.

In the page on the right
A Two-ship formation approaches the rugged California coastline.

In the page overleaf
Two Vipers turn over the Ocean on CAP assignment in the San Francisco area.

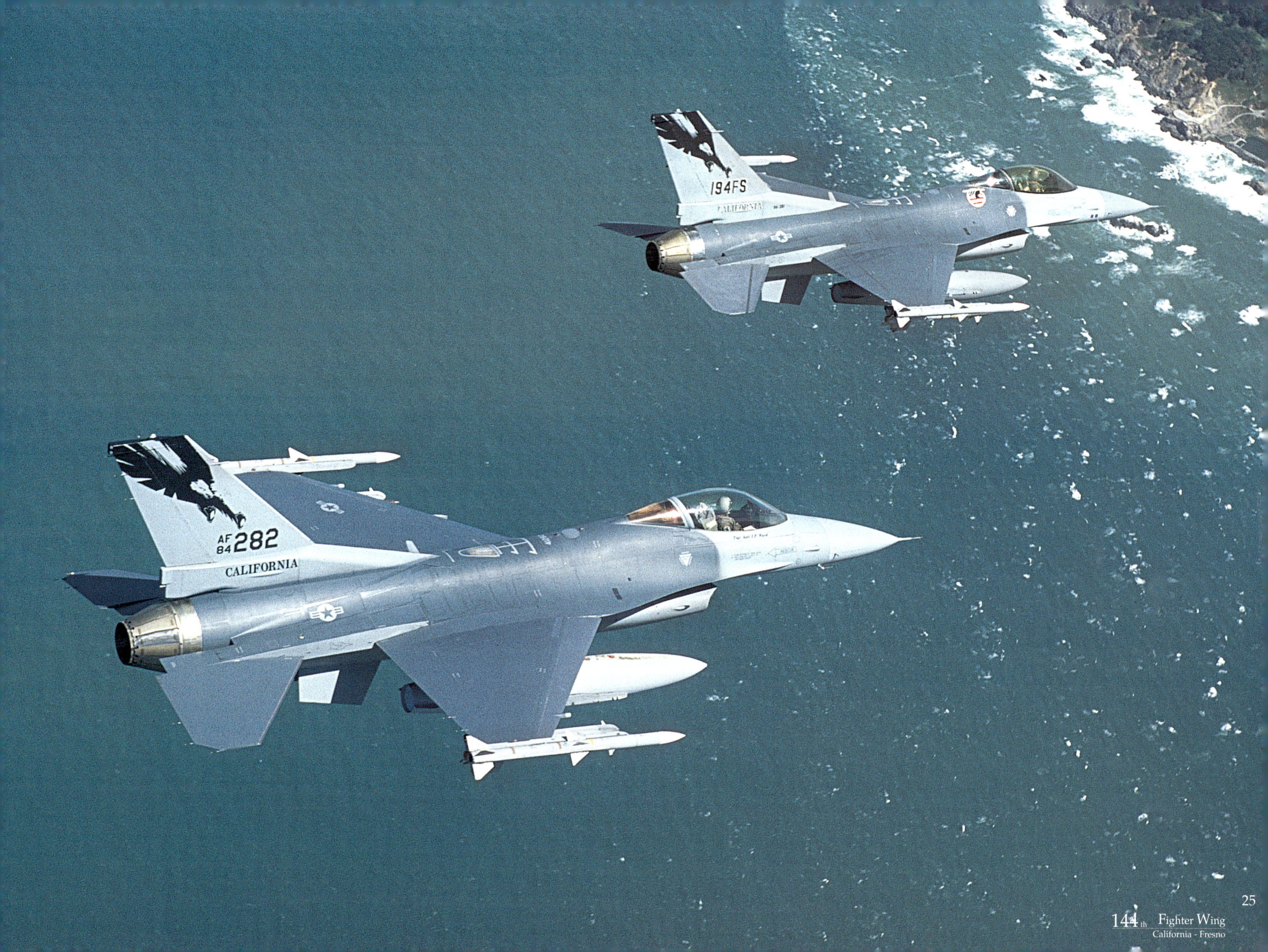
194FS
CALIFORNIA
AF
84
282
CALIFORNIA

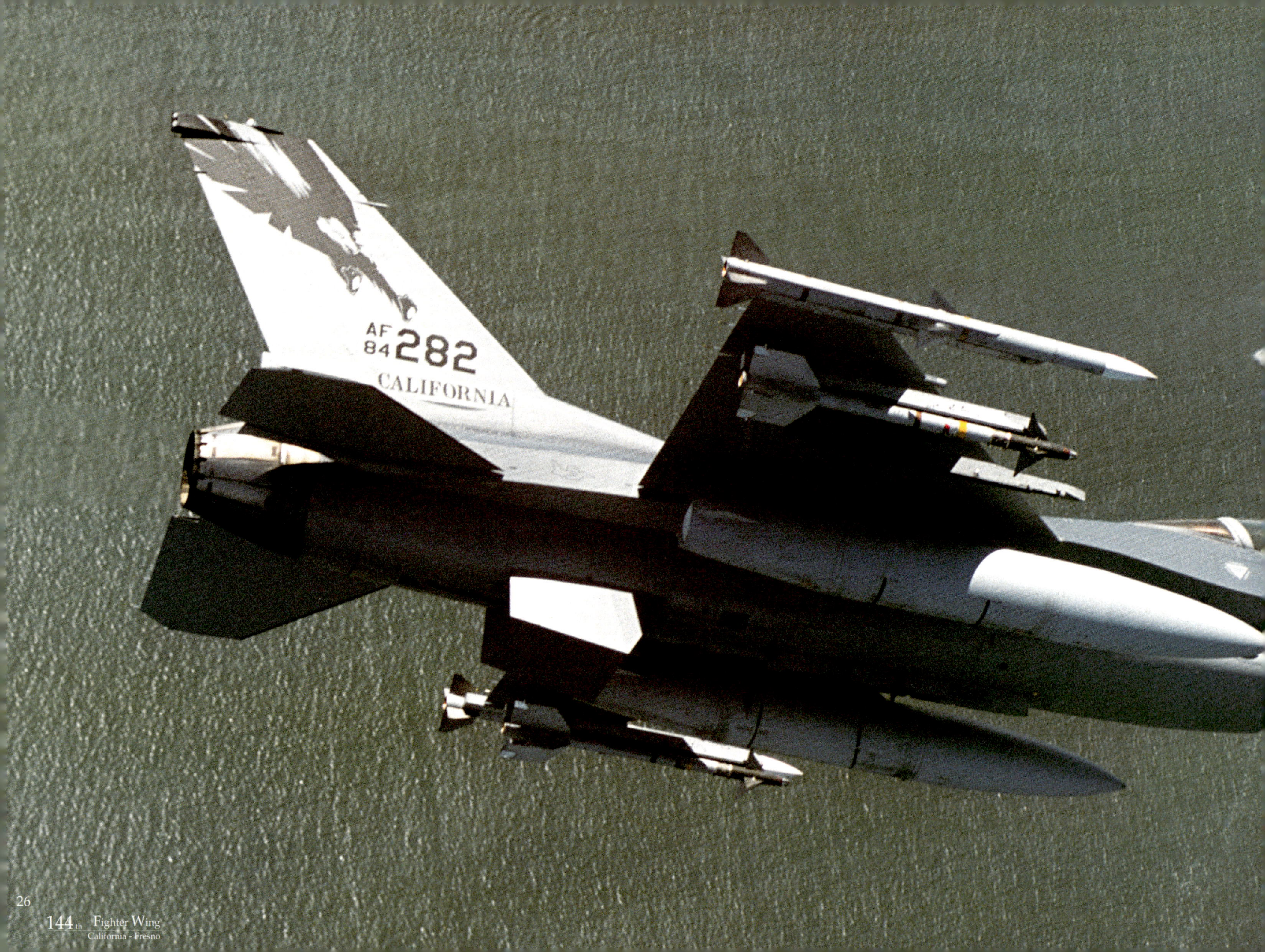
AF
84 282
CALIFORNIA

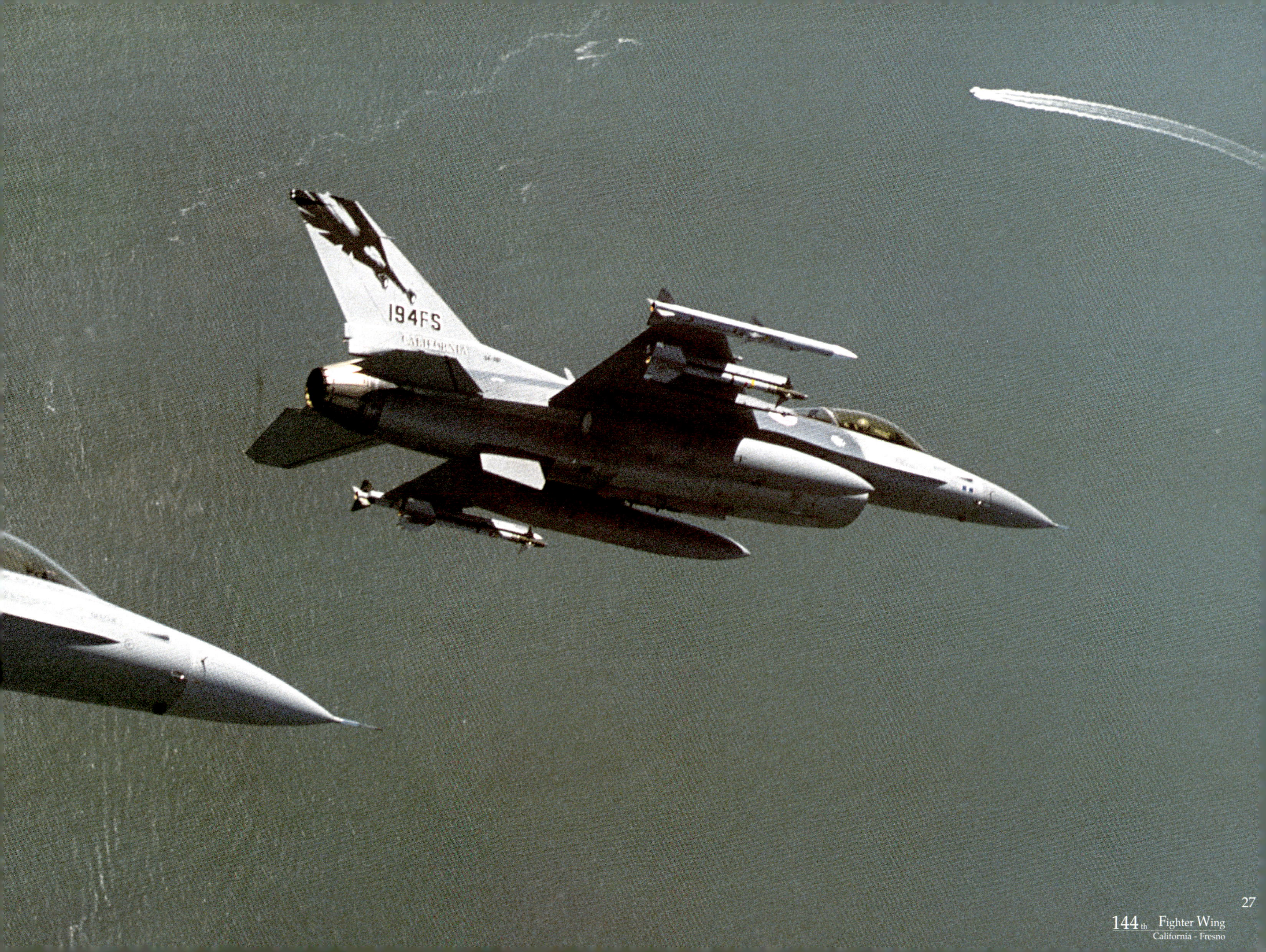
194FS
CALIFORNIA

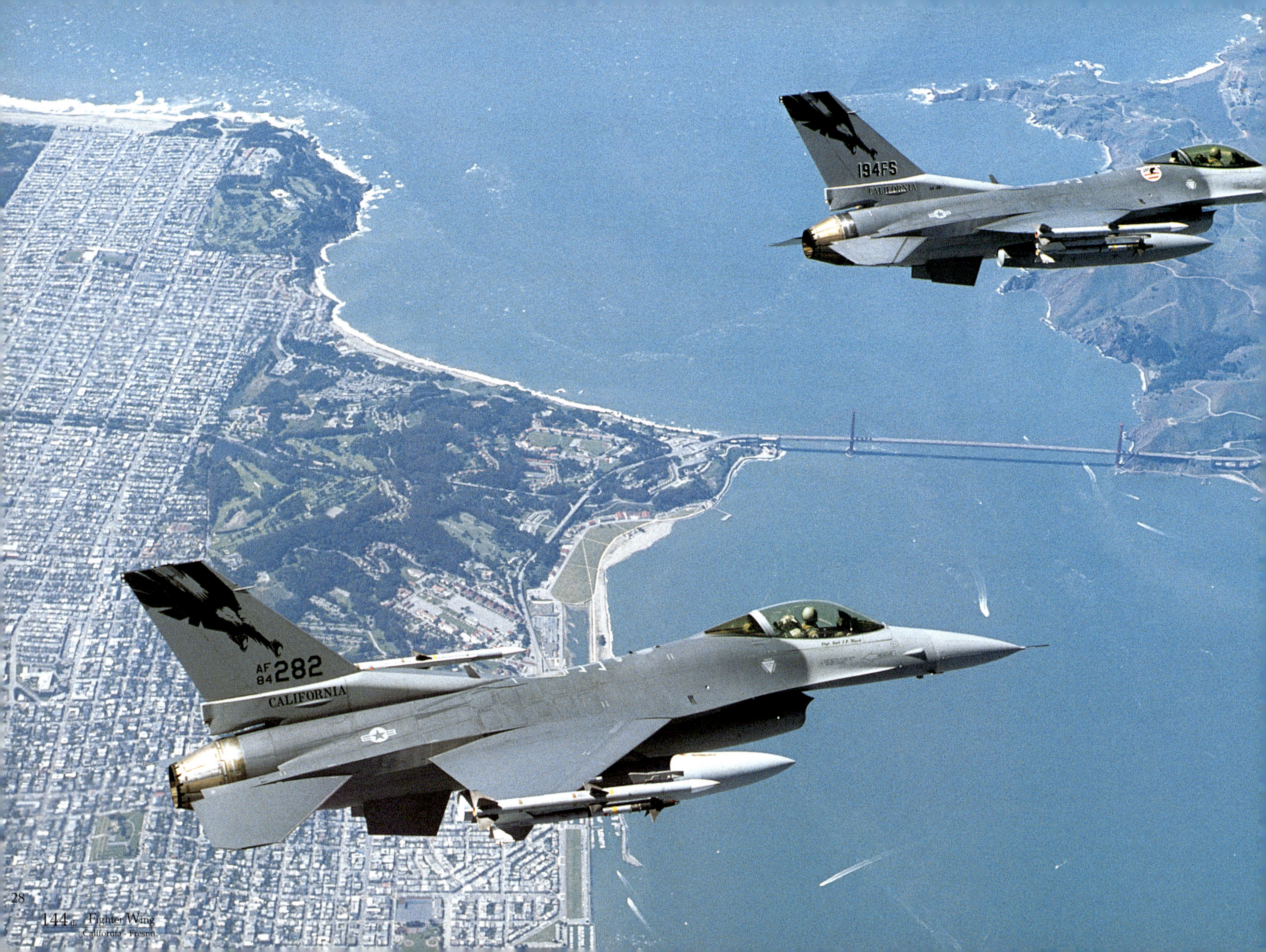
194FS
CALIFORNIA
AF 84 282
CALIFORNIA

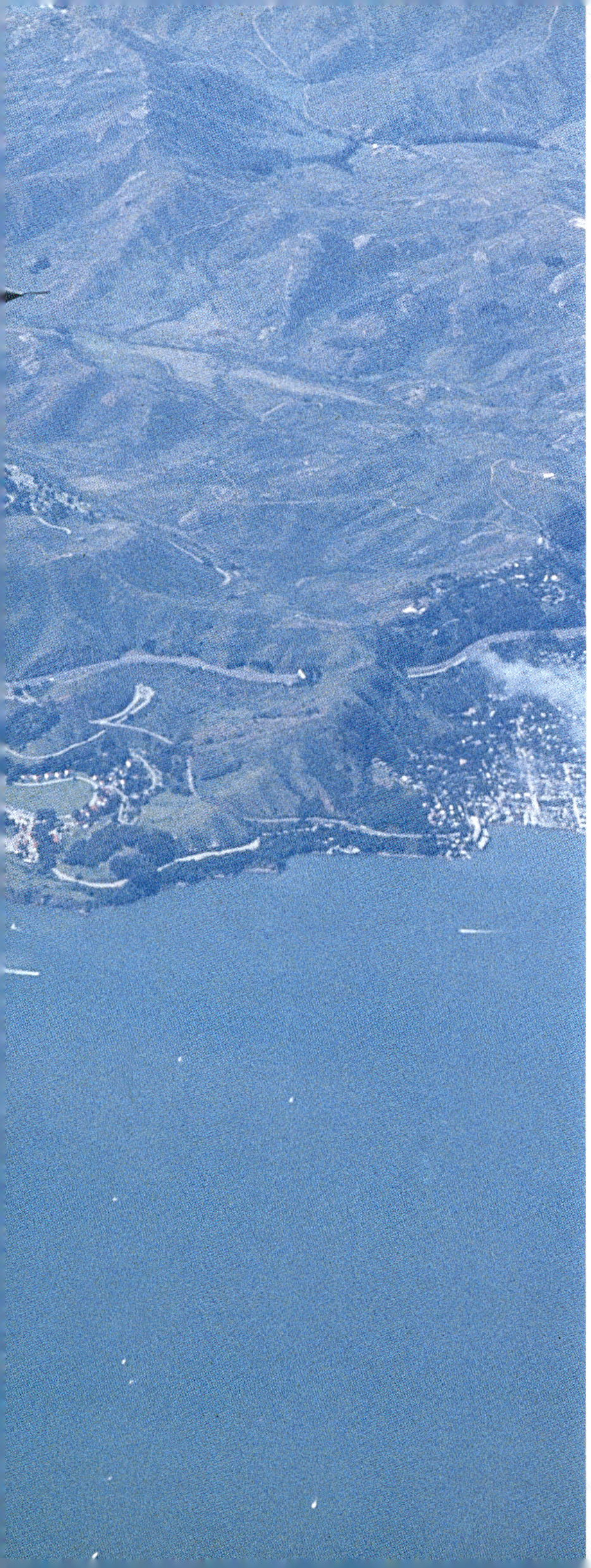

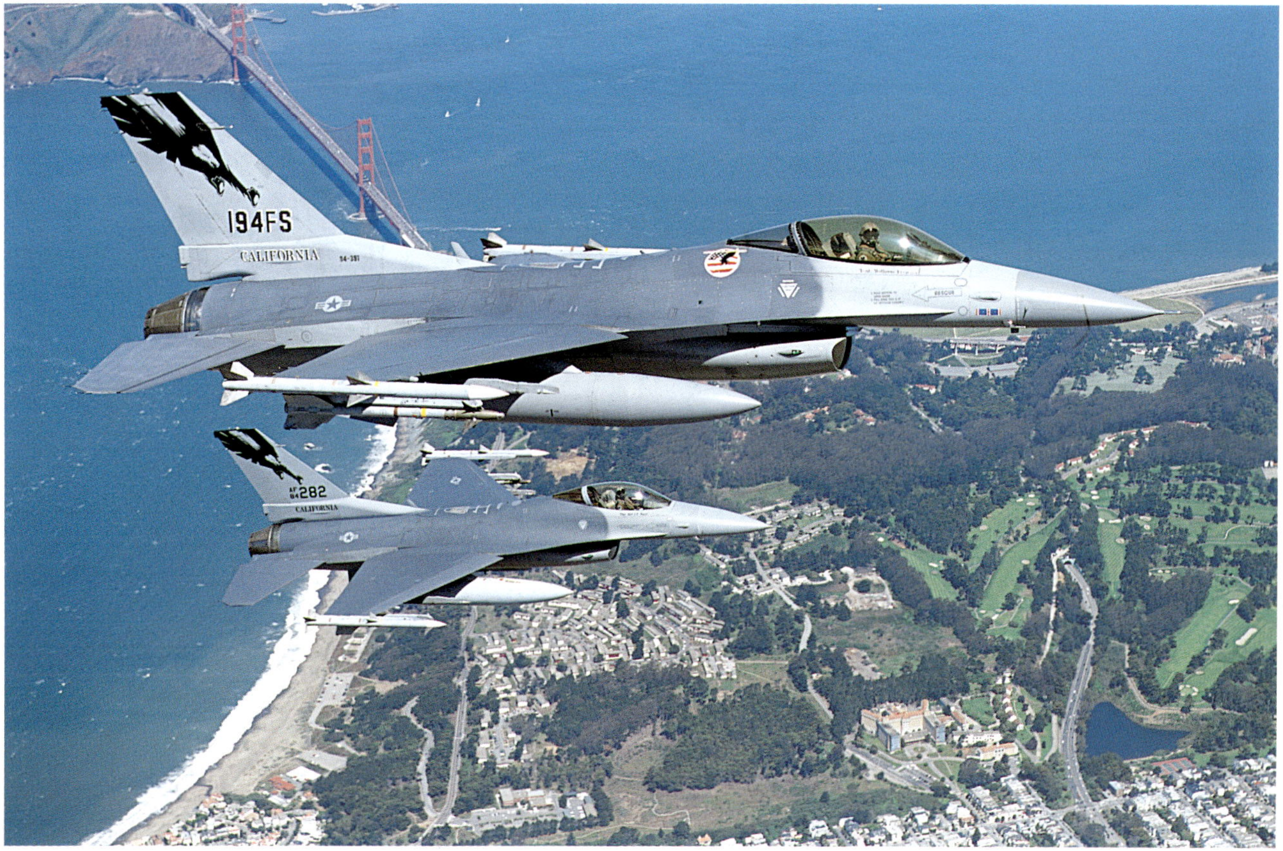

Above
A two-ship formation of 194th Fighter Squadron on an easterly heading to overfly San Francisco.

On the left
Flying high over SFO area, the Golden Gate bridge is a good visual waypoint.

So, are your "Griffins" of the 194th Fighter Squadron a benchmark organization? You bet they are! Teamwork in the air and on the ground has brought the "Griffins" to this high level of performance and professionalism.

In today's rapidly changing military environment, only the fighter squadron who can adapt quick and often will succeed. We are that squadron.

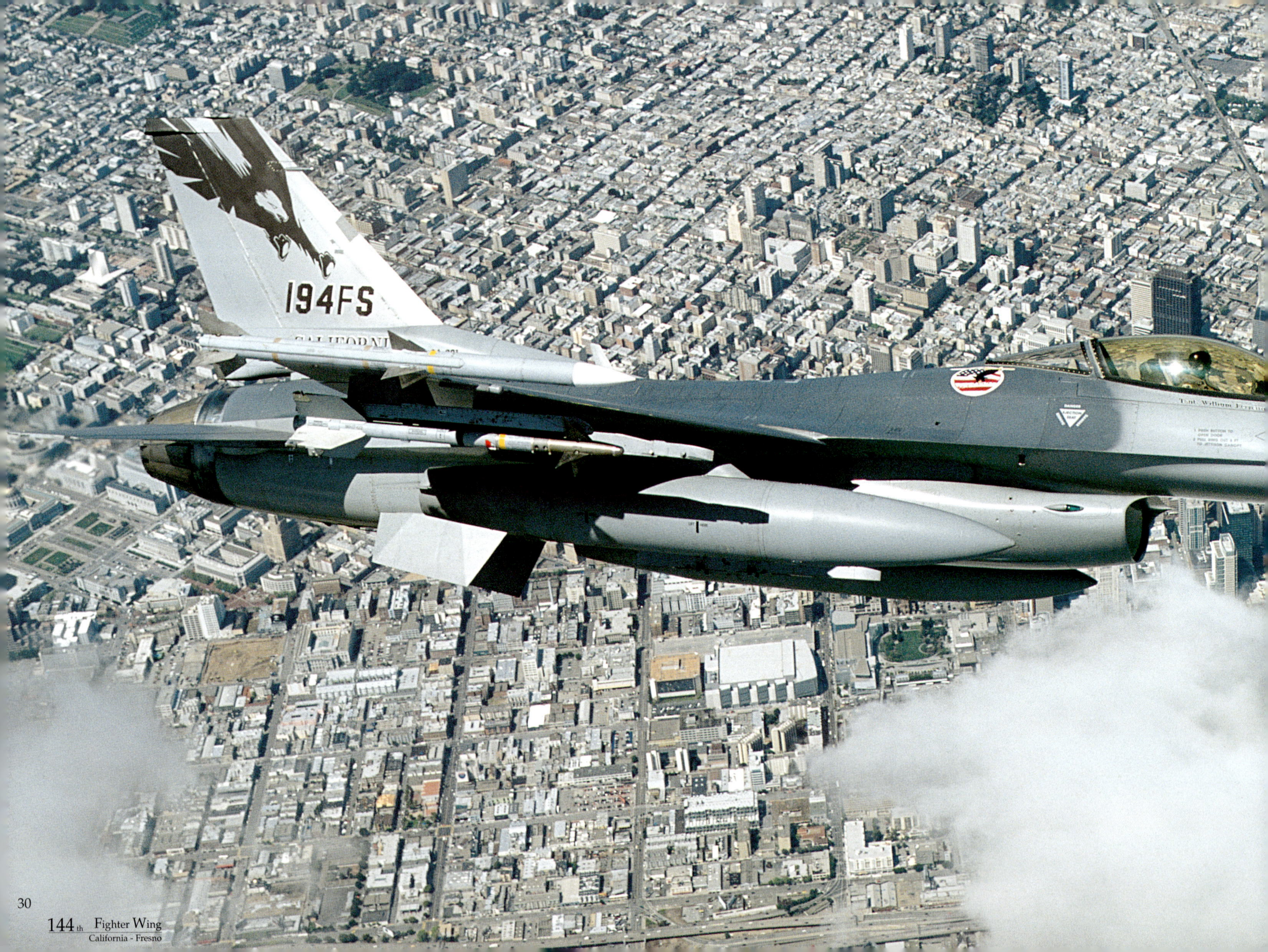
194FS

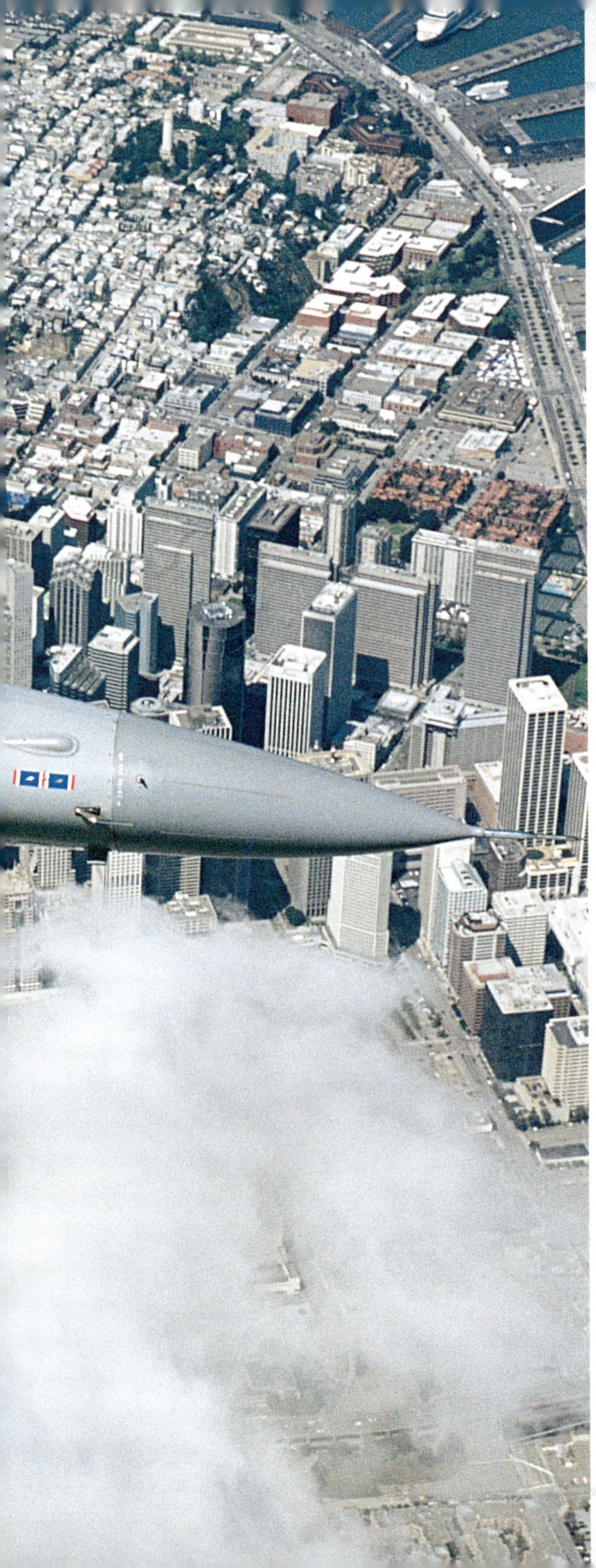

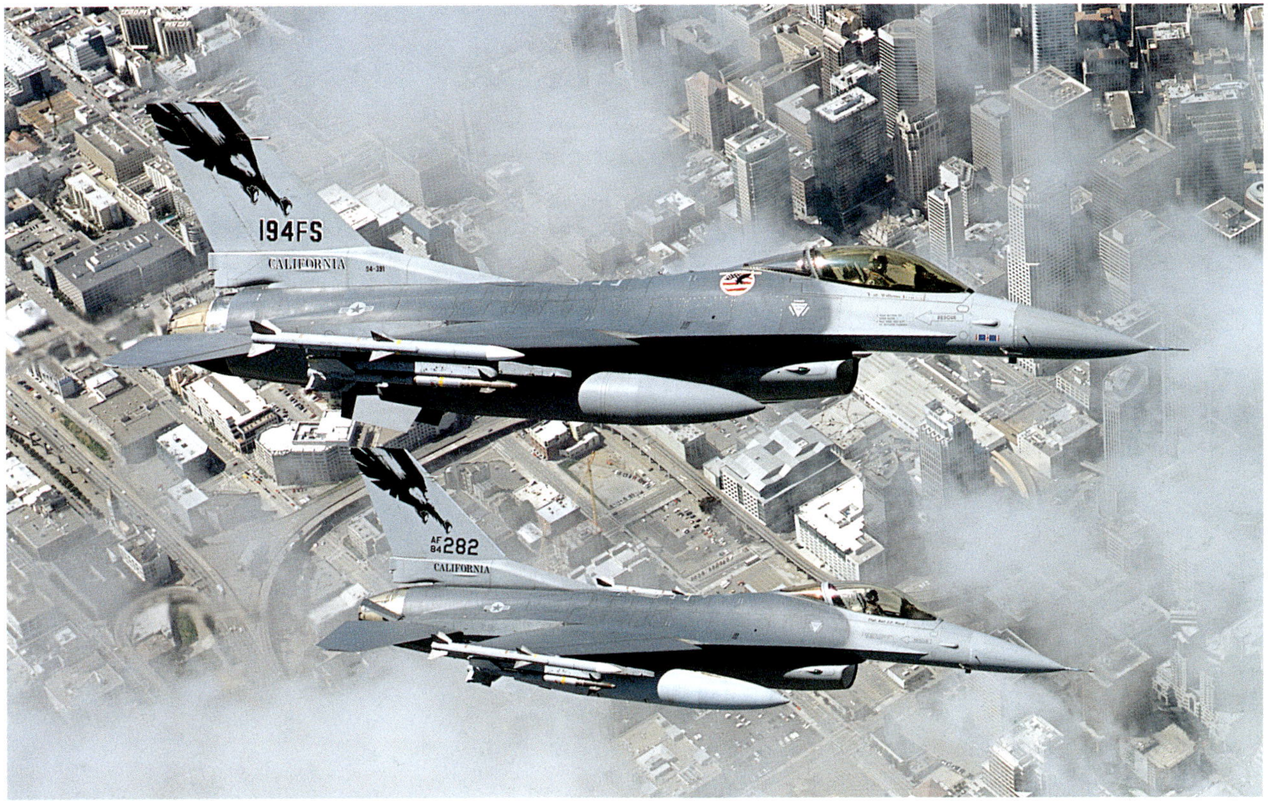

Above
A 144th Fighter Wing CAP over downtown San Francisco.

On the left
Several millions of inhabitans live safely under the wings of this 144th FW Viper.

On the right
An unofficial patch worn by California ANG crews.

In the page overleaf
The formation splits to proceed on CAP at lower altitude.

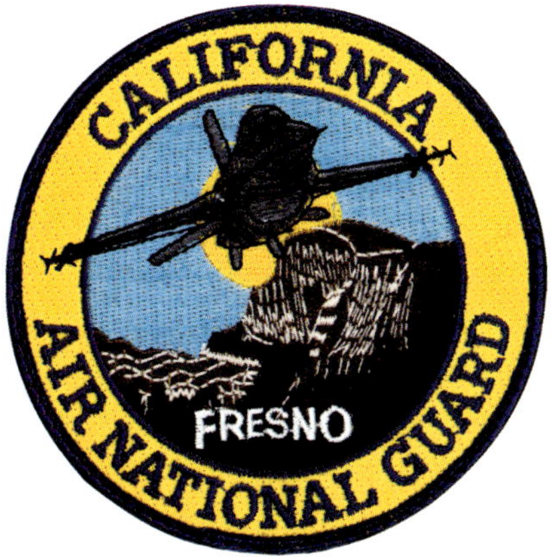

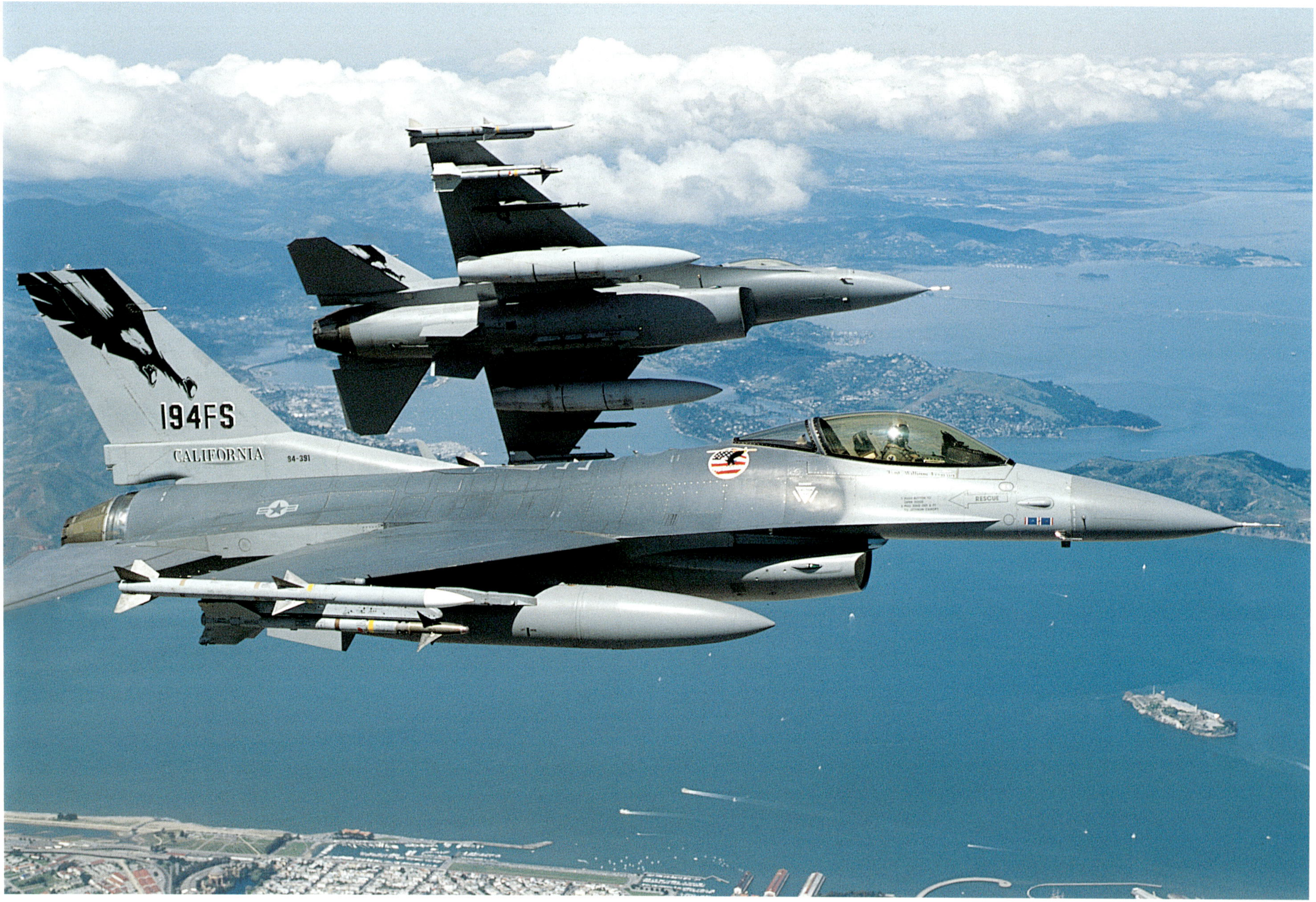
194FS
CALIFORNIA
84-391
RESCUE

114th Fighter Wing

South Dakota

114th Fighter Wing, South Dakota (emblem approved 3 August 1965). The thirteen stars represent the number of the original U.S. Colonies and the thirteen officers who established the South Dakota ANG. The flaming sun represents the thrust and power of the aircraft. The radar scope and lightning bolt represent the use of radar and the all-weather flying ability. The contrail ending in a falcon symbolizes retaliation and swiftness of action. The emblem contains Air Force yellow and blue and the national colors of red, white, and blue.

175th Fighter Squadron, South Dakota (emblem approved 9 February 1949). The "lobo" or coyote is a western animal native to South Dakota, which is also known as the "Coyote State". The coyote is used in the heraldry of the state and its National Guard units. The fighter squadron, like the intelligent and ferocious coyote, is ready at all times to guard, defend, and attack whenever necessary.

175th Fighter Squadron, South Dakota "Lobos". Constitued at Richmond Army Air Base in April 1943 as the 387th Fighter Squadron. Assigned to the 365th Fighter Group, 9th Air Force, and equipped with Republic P-47D Thunderbolts. The squadron flew combat missions in the European Theater of Operations (ETO) from February 1944 until Victory Europe (VE) Day, 8 April, 1945. The unit was equipped with North American P-51D Mustangs and mobilized for the Korean War on 1 March 1951. The squadron was assigned air defence duty at Ellsworth AFB, South Dakota. The unit returned to state control on 1 December 1952 at Sioux Falls Municipal Airport. The 114th Fighter Group was activated on 16 April 1956 as a headquarters element for the squadron. Since June 1954 the squadron has flown the following types of aircraft: Lockheed F-94A/B/C Starfire, Northrop F-89D/J Scorpion, Convair F/TF-102A Delta Dagger, North American F-100D/F Super Sabre, Vought A-7D/K Corsair II, and in 1992 the General Dynamics

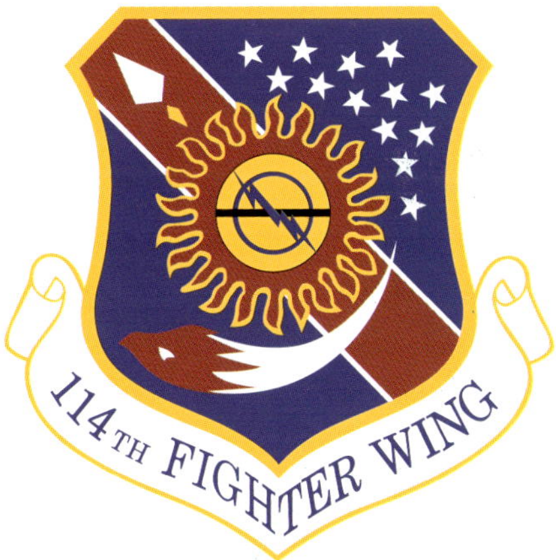

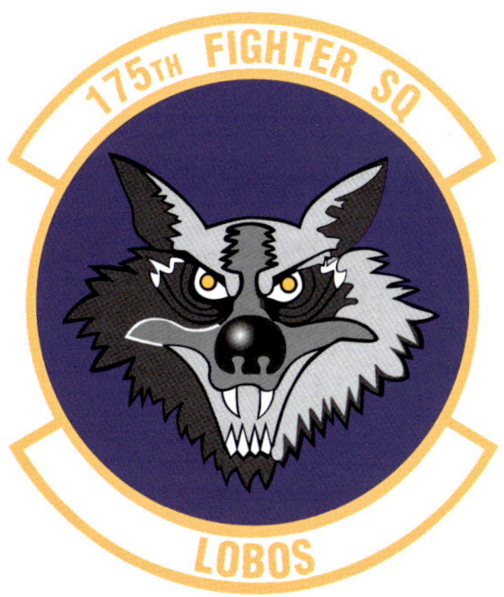

(then Lockheed Martin) F-16C/D Fighting Falcon. The unit served an air defense role until Spring 1970 when with F-100Ds it changed missions to become a tactical fighter unit. In 1989 the unit flew combat missions with A-7Ds in Operation *Just Cause* during the invasion of Panama.

"LOBO 11 IS IN FROM THE WEST"

BY LT.COL. ROD RODRIGUEZ

It's 7:00 AM. I just finished breakfast at KC's restaurant, and the weather is clear and brisk. The temperature is 8 °F, but at least the winds are calm. It's January in Sioux Falls, South Dakota, a sunny day with single digit temperatures and calm winds means pretty good flying weather.

The schedule in Ops has me leading a four-ship of F-16s to Smokey Hill bombing range in Kansas. The rest of the four-ship consists of Lumpy,

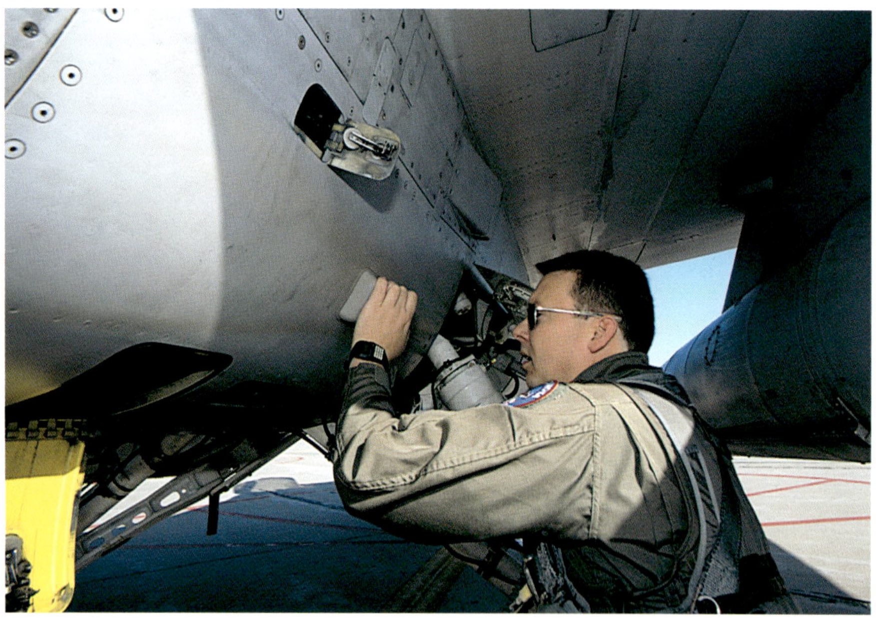

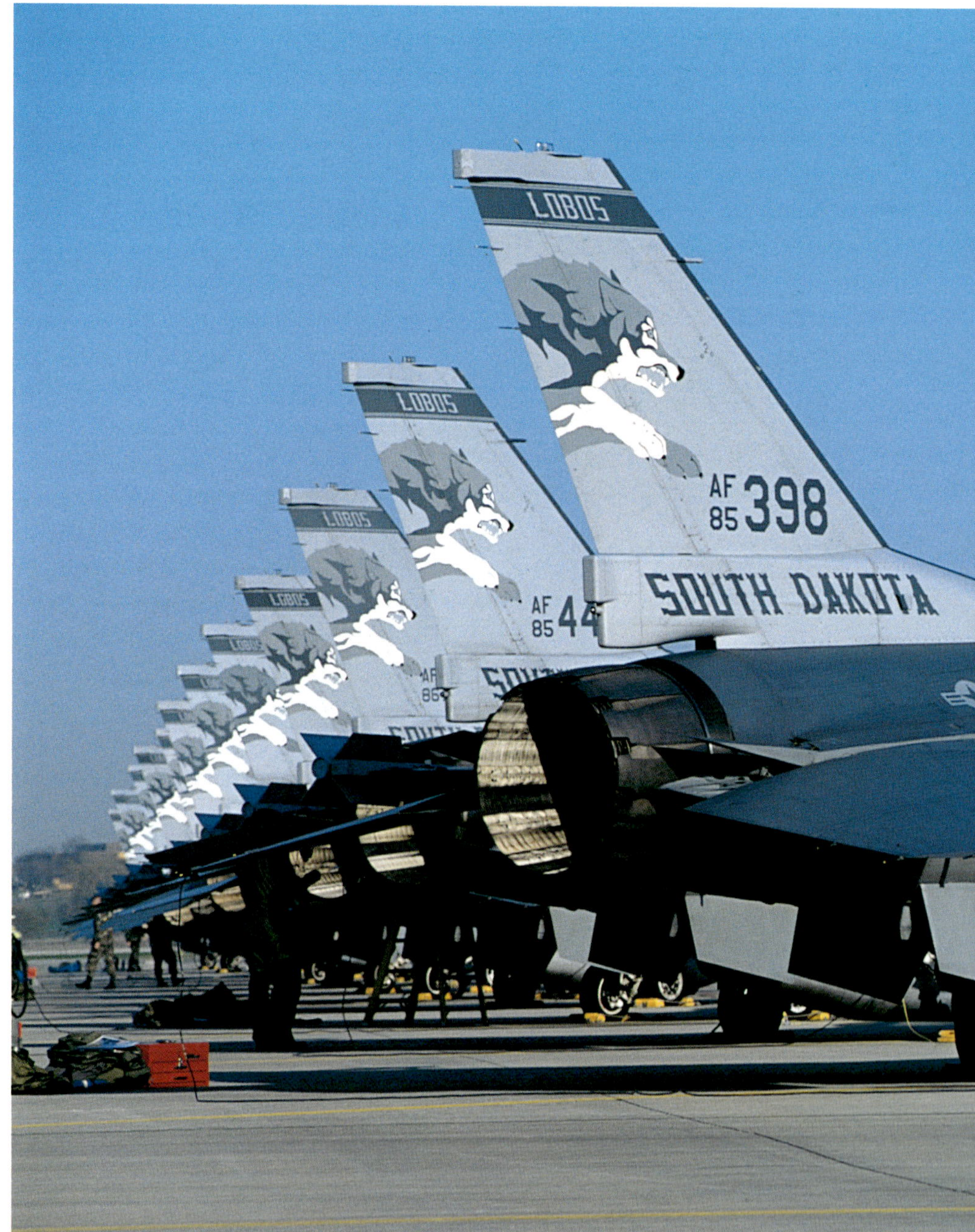

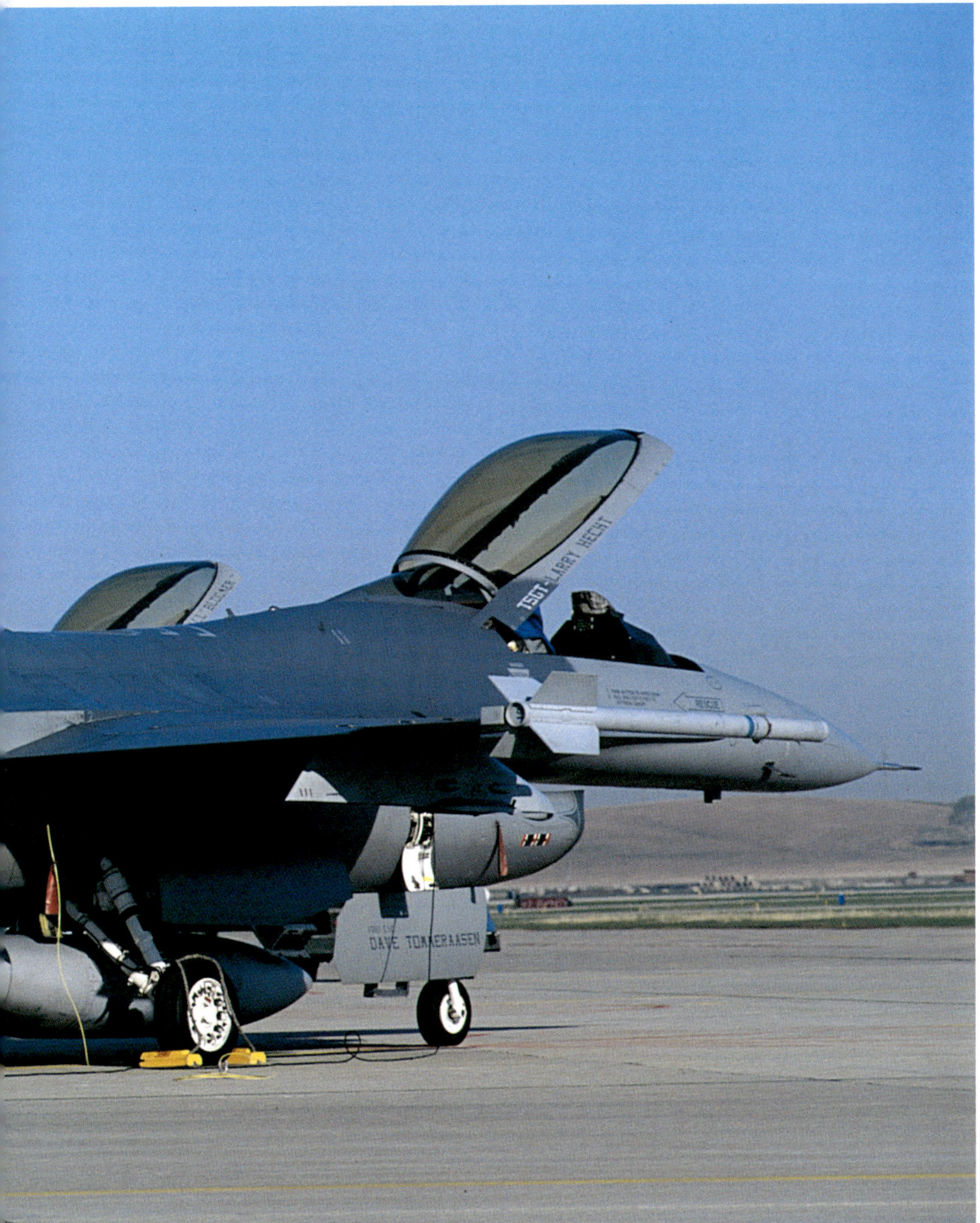

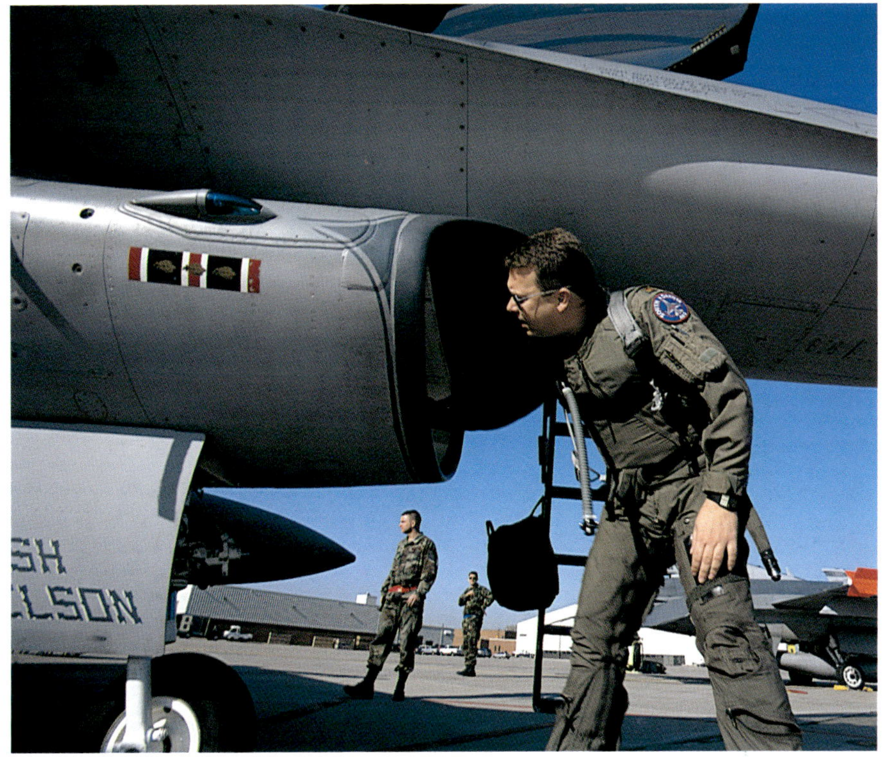

Above
Before getting into the cockpit, the pilot performs a series of pre-defined checks.

On the left
All the F-16Cs Block 30 of the 175th Fighter Squadron are perfectly lined-up on the air base flight line.

In the page on the left
A "Lobo" pilot busy with his pre-flight check.

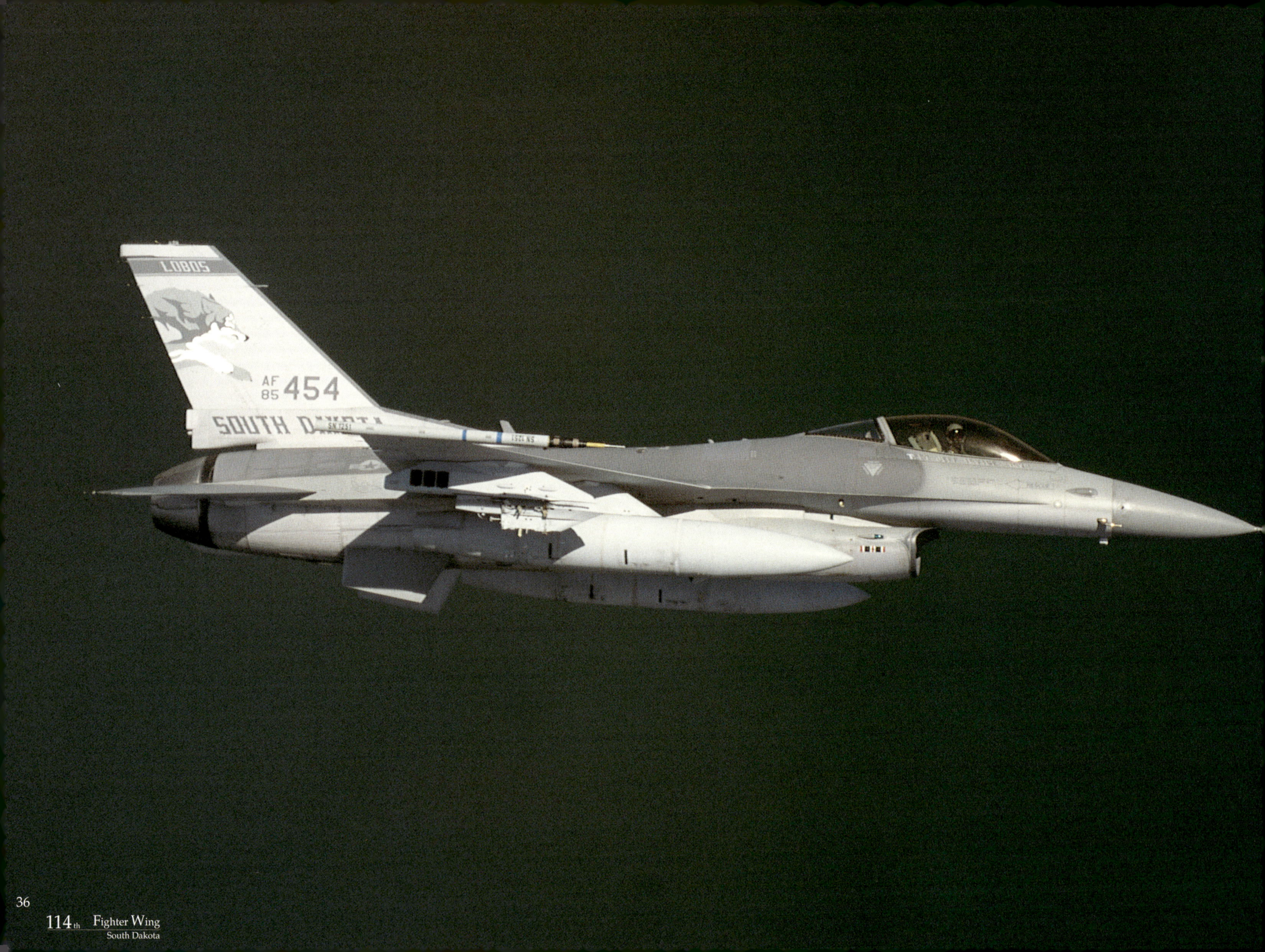
LOBOS
AF
85 454
SOUTH DAKOTA

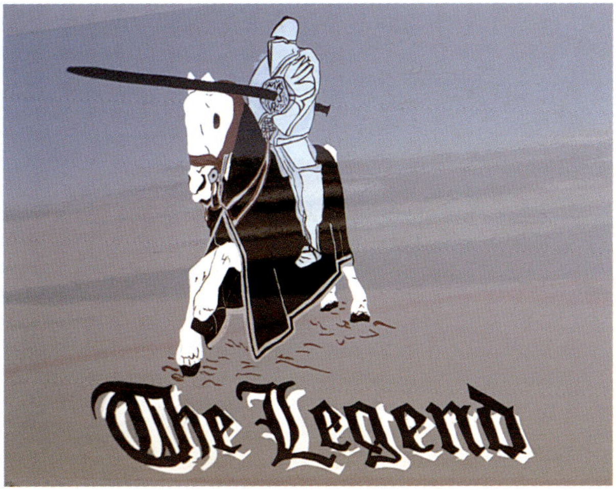

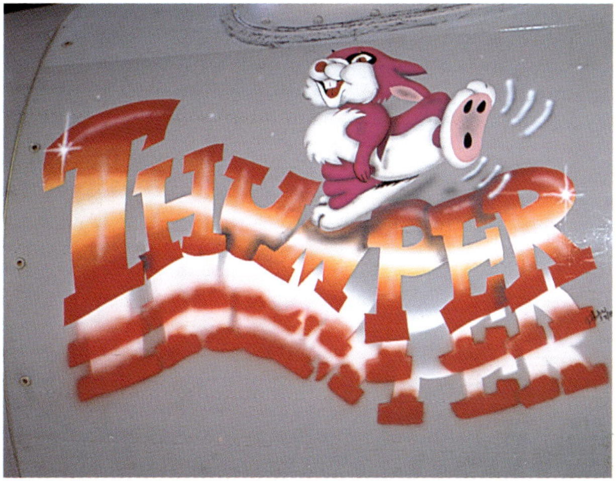

Above and on the right
Painting artwork on their own aircraft is a long standing tradition at 114th Fighter Wing.

On the left
Lt.Col. Dick Rayburn, 114th FW Chief of Operations, turns in his Viper over a lake formed by a dam along the Missouri River.

Chairman, and Eddie (our new lieutenant). Flight planning goes quickly, as all four of us work to prepare materials and weaponeering. We usually brief between 1 and 2 hours before take-off. Since Eddie is fresh out of initial training, I plan to use the full 2 hours to make sure all of his questions are answered and he is clear on his duties during the flight. The briefing goes smoothly. "The bet is standard", twenty-five cents per bomb, and winner takes all. Eddie is progressing well.

He will mature into an outstanding fighter pilot for the "Lobos". After briefing each element of this morning's sortie, we collect our winter flight gear from life-support. Then it's one last stop at the bathroom and then at the Ops desk for our 8-mm

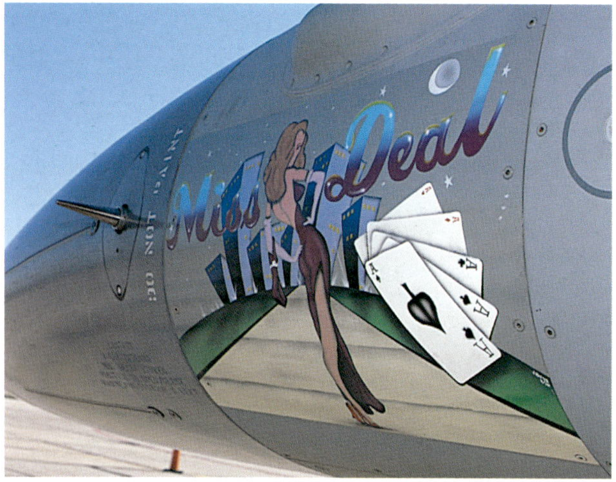

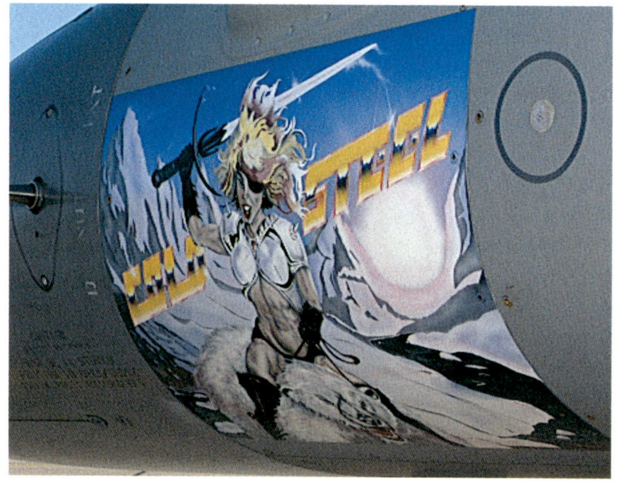

tape and assorted in-flight classifieds... out to our single-seat, single-engine fighters. If you don't think there's a difference... you just don't get it. The step and start proceed quietly. I've flown with many active duty and ANG F-16 units; "Lobo" maintenance is second to none. Just before we taxi, I establish communications by "checking in" the flight on our UHF and VHF radios. Next we receive our flight clearance from the control tower. A quick taxi to the arming area allows our weapons loaders to arm our bombs and 20-mm Gatling gun. Today we are carrying 6 BDU-33 practice bombs and 20-mm target practice rounds. BDU-33s are 25-lb. bombs with a phosphorous charge that allow us to see the point of impact in relation to our intended target more

Above and on the left
The "nose art" has always been an important communication component of US Air Forces.
At 114th FW, it shows up on each Viper left side, just past the radome.

On the right
In a low-level mission, the never-ending South Dakota plains streak under the wings of the Viper

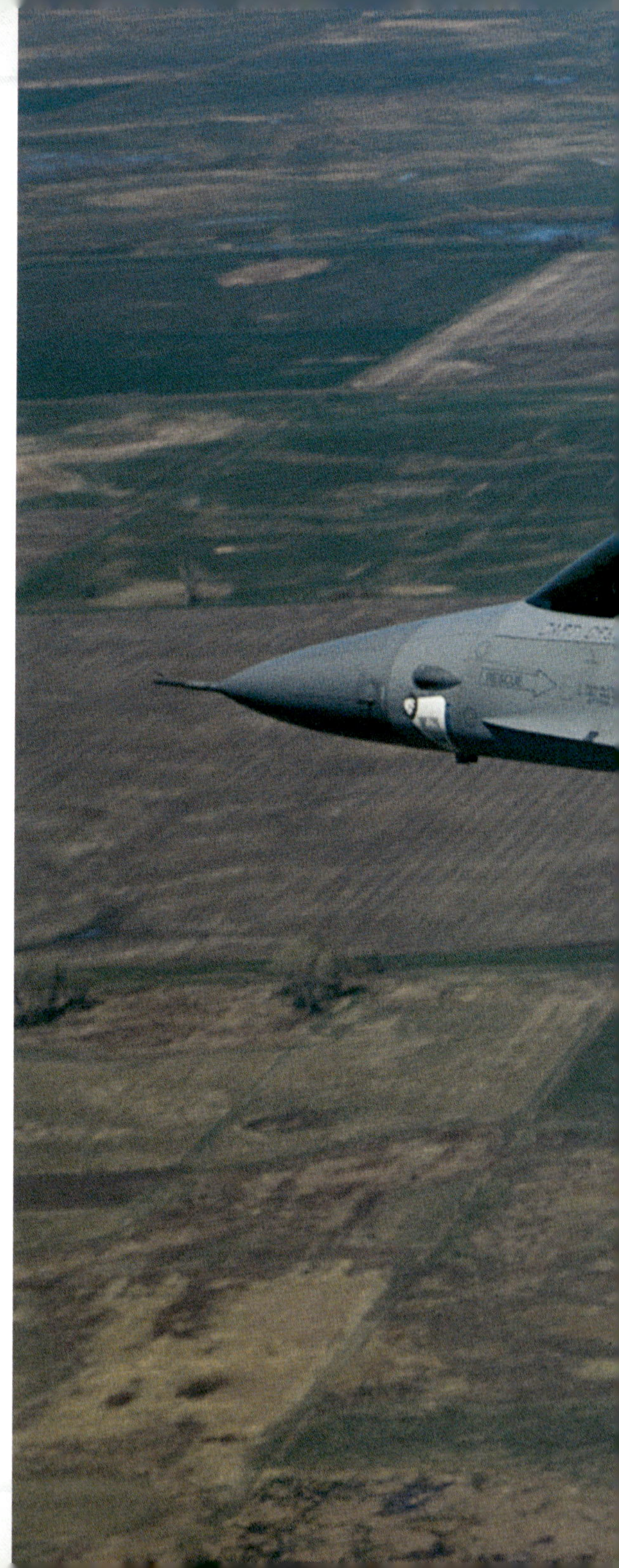

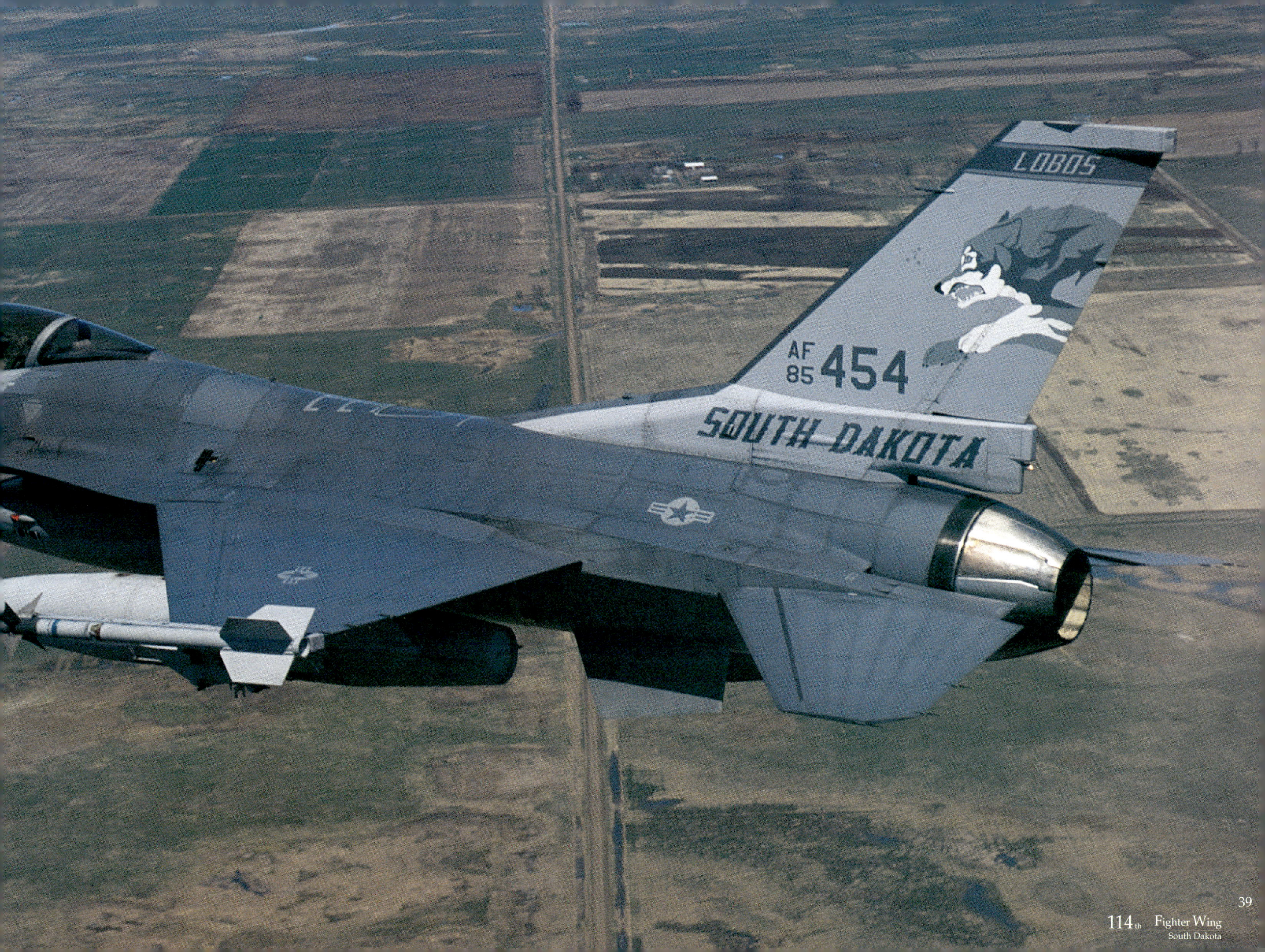
LOBOS
AF
85 454
SOUTH DAKOTA

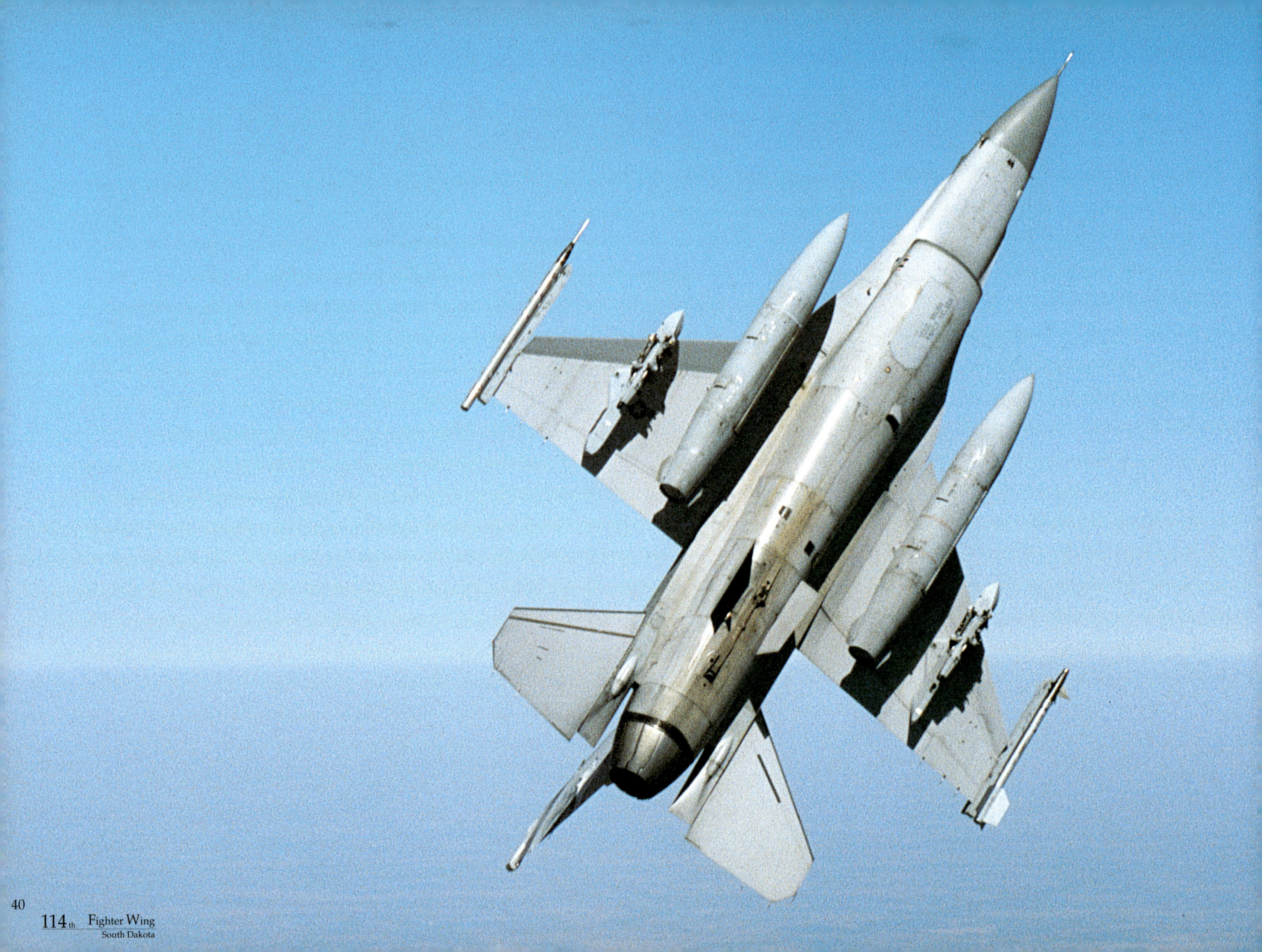

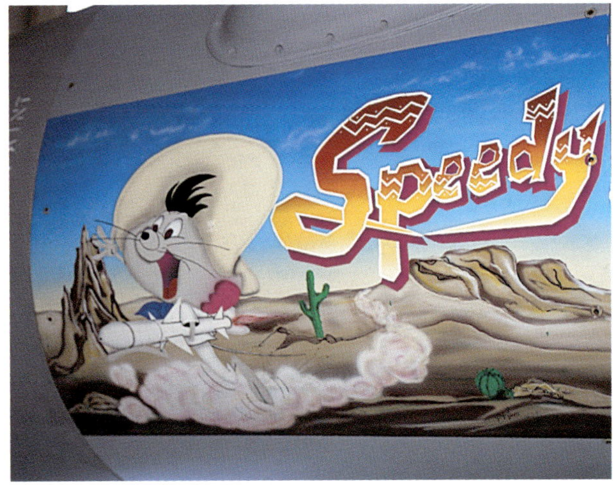

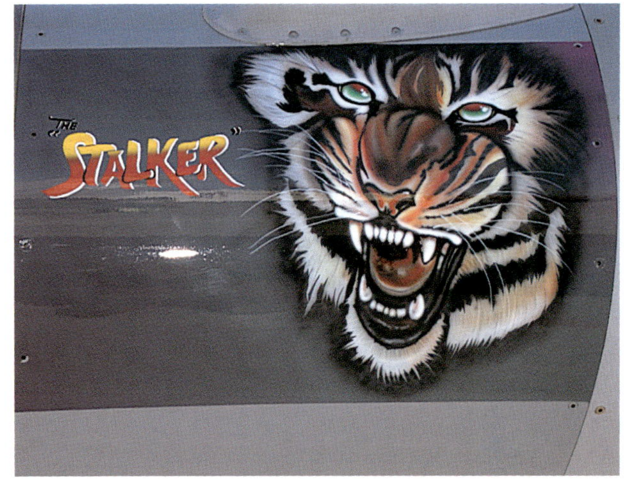

Above and on the right
The 114th FW "nose arts" are painted with precise craftmanship. An aerograph pen is commonally used.

On the left
A 114th F-16C pops up nicely in a spot of the North Dakota blue sky.

clearly from the air. We use the BDUs to simulate a variety of live munitions; they're much cheaper and have flight characteristics similar to many bombs in our inventory. After the jets have been "armed up", we head for the runway.

Once we receive take-off clearance, all four of the jets taxi onto the runway. When everyone is in place, packed closely together with only wingtip separation, I give the "run it up" signal and all the jets strain against their brakes... We're ready to go.

The noise outside our jets is deafening. After the pilots have checked over their jet one last time, I get three "thumbs up" from the guys. When I release the brakes, I select afterburner and my jet accelerates quickly. At 100 knots, I disengage the nose-wheel

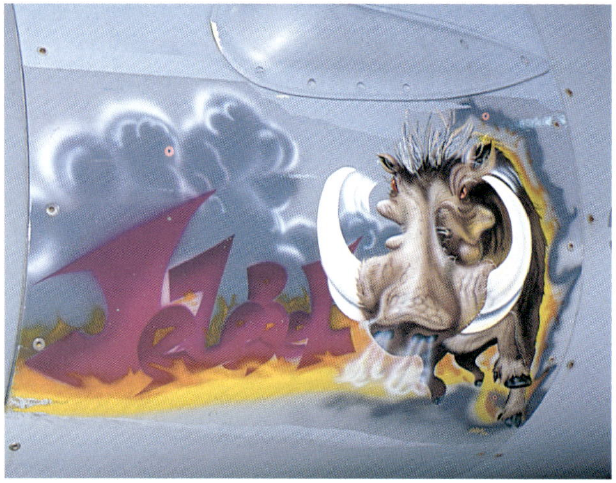

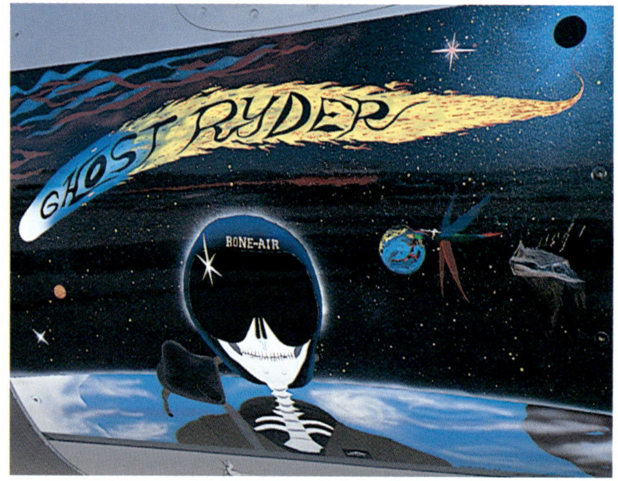

steering. Shortly after that, I pull back on the stick and rotate the jet for take-off. At 165 knots and about 2000', I'm airborne. Everything looks good, and the landing gear is coming up. Twenty seconds after I released my brakes, number 2 releases his ones. He is followed, 20 seconds later, by number 3 and then number 4. Sioux Falls Departure Control quickly clears us on course, then hands us off to Minneapolis Center.

The distance to Smokey Hill is approximately 300 nautical miles. At the speed we fly it should take about 40 minutes. Enroute we reform into a fighting formation and practice visual lookout, radar sorting/targeting and communications. About 100 miles from the range we descend into low level training

Above and on the left
When a "nose art" is superseded by an other, then it is reproduced on a wall in the 114th Fighter Wing meeting room.

On the right
A 114th Fighter Wing "Lobo" in right-wing station during a tactical flight.

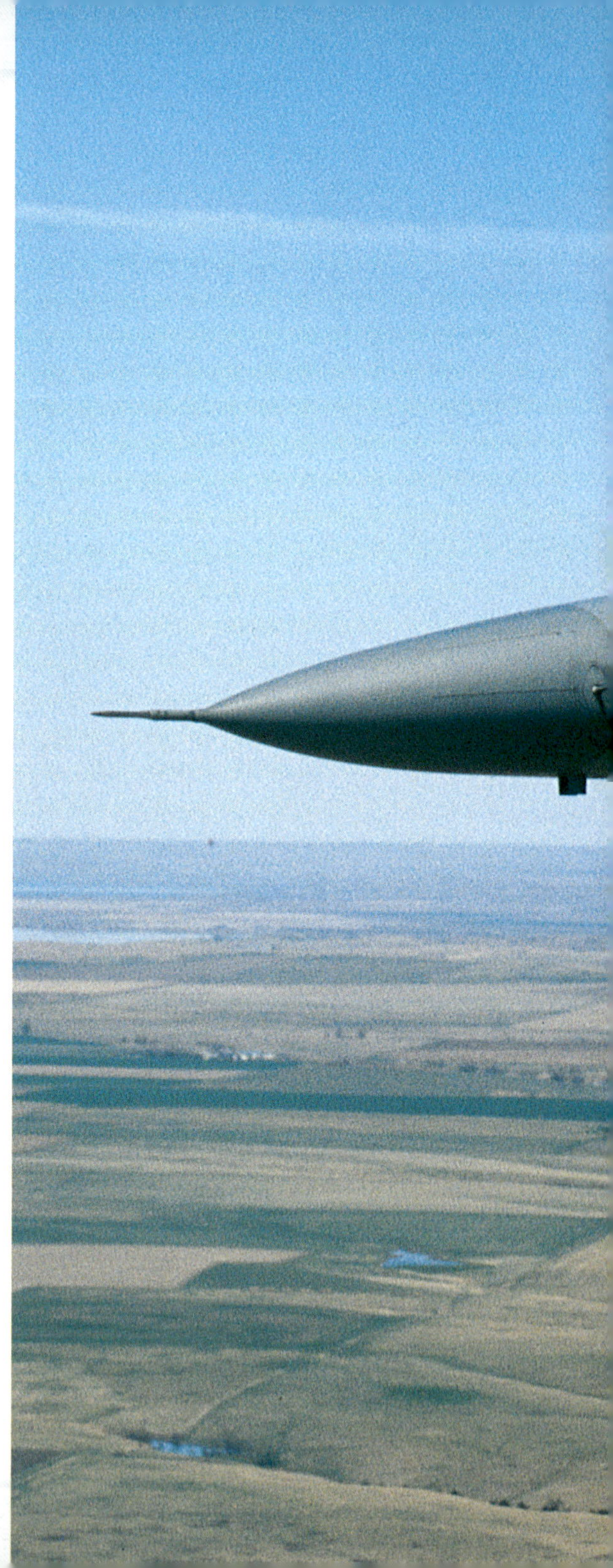

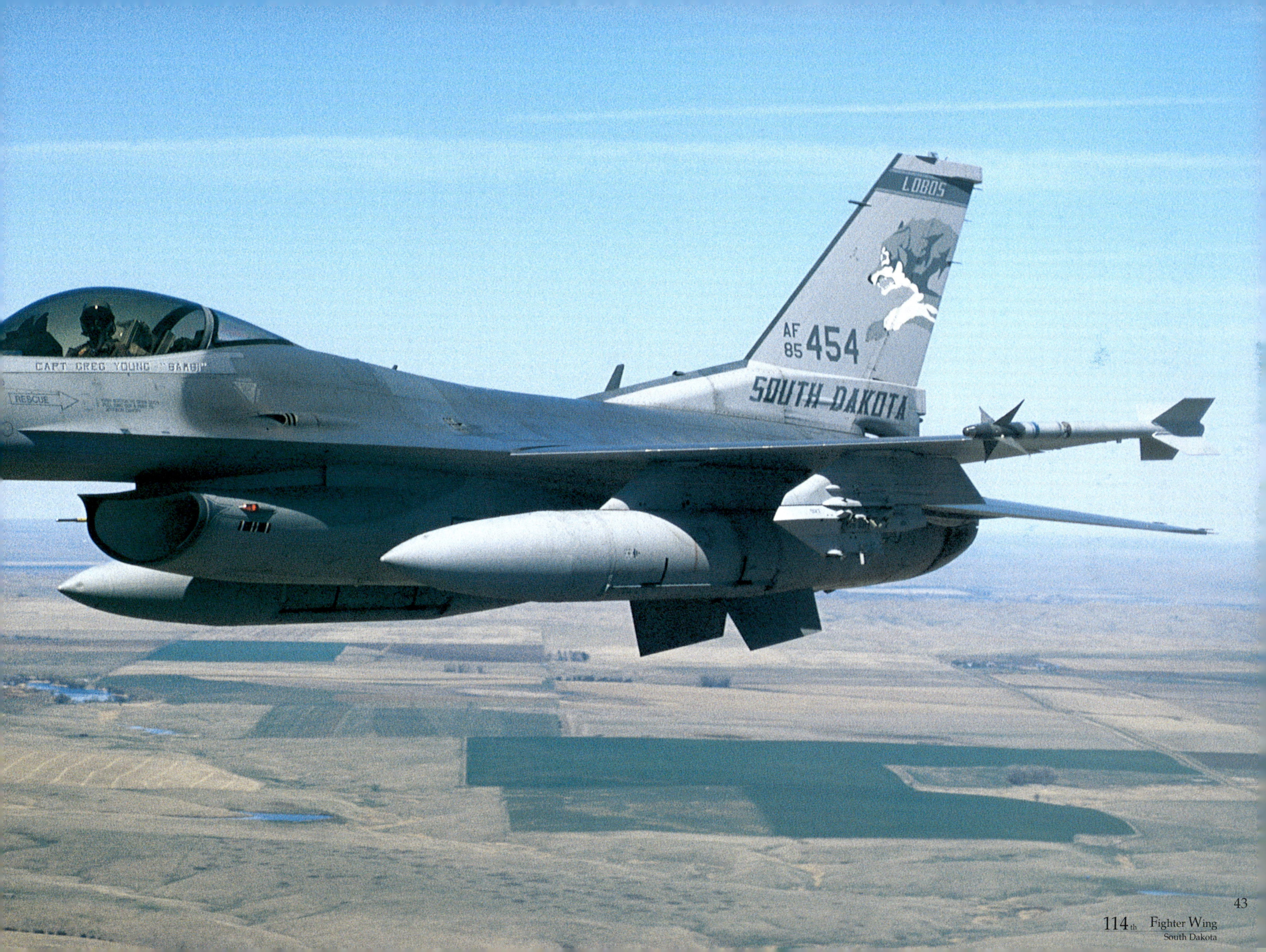
LOBOS
AF
85
454
SOUTH DAKOTA
CAPT GREG YOUNG "BAMBI"
RESCUE

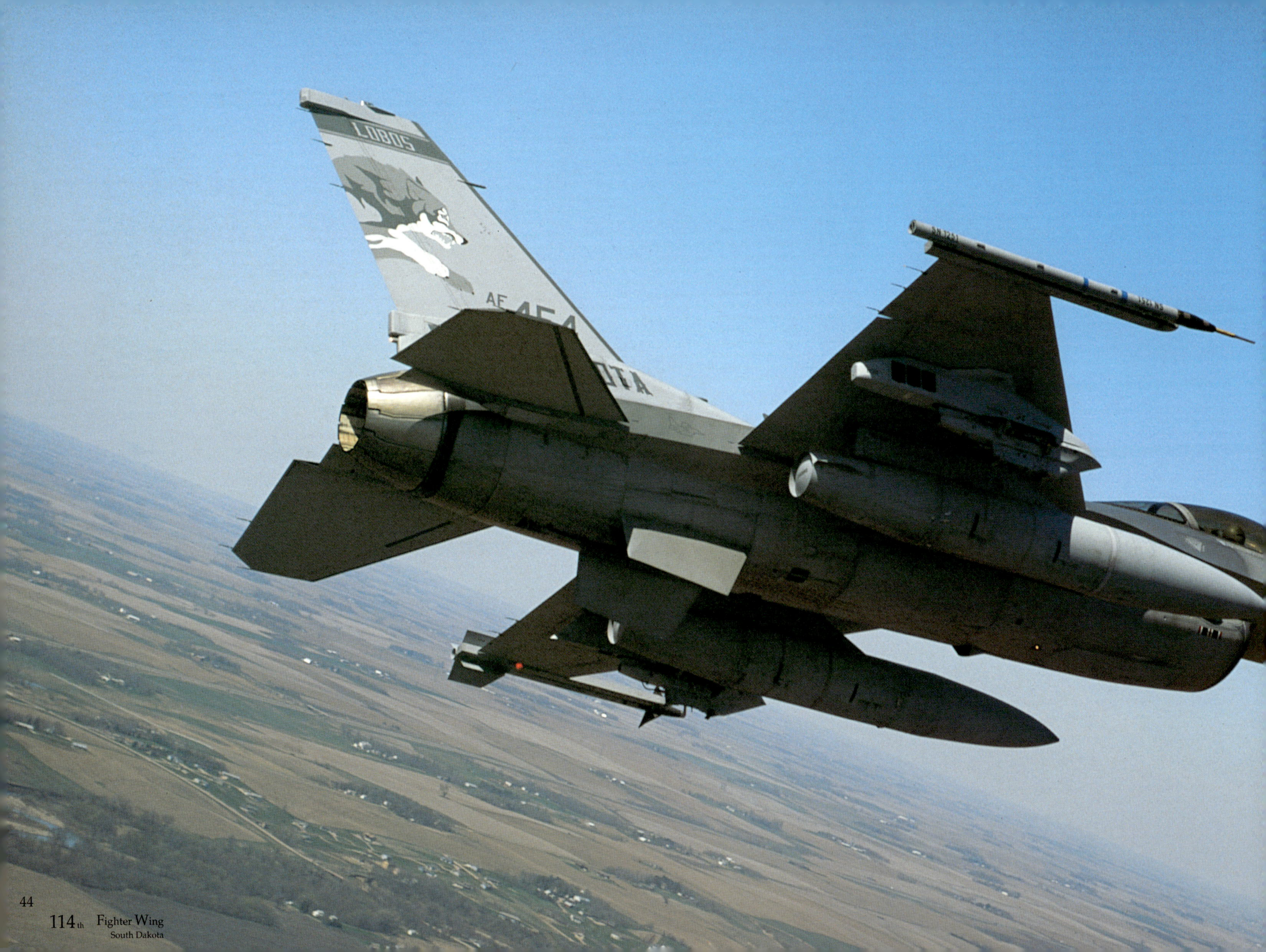
LOBOS
AF

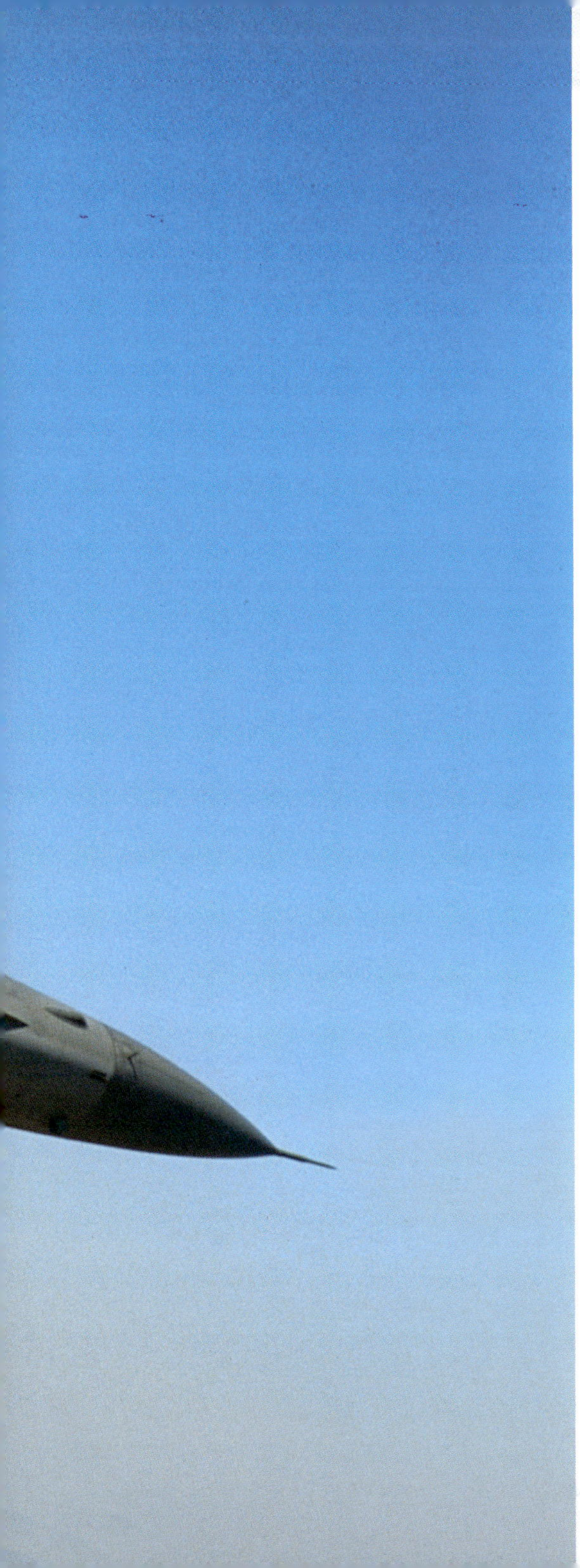

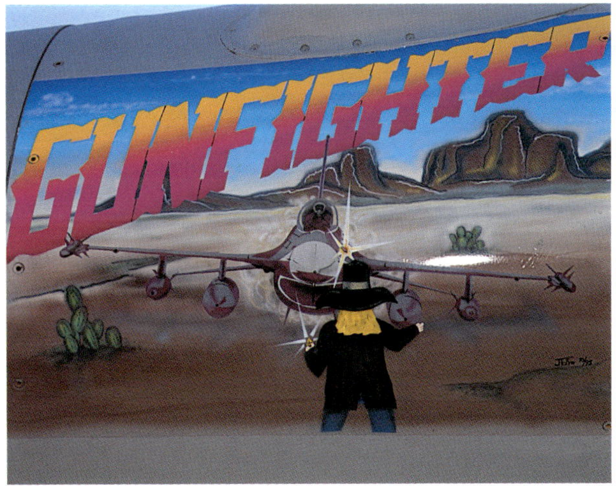

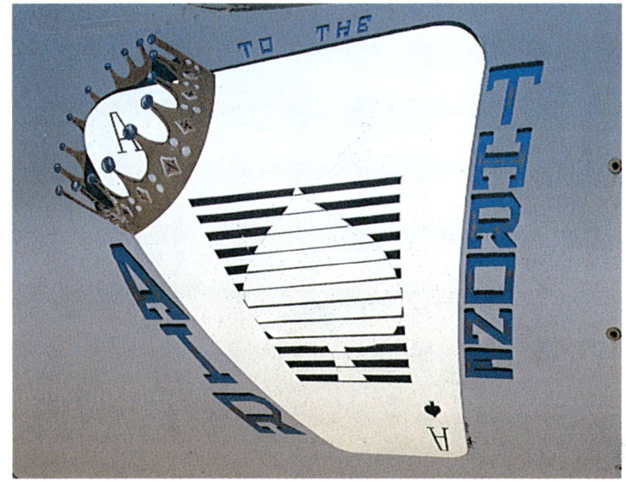

Above and on the right
The "nose art" is often definitely a very nice and colourful touch on a Viper's side.

On the left
The pure clean shape of the Viper shows off very well in this air-to-air shot.

area that leads to the range. As we enter into the training area, we accomplish a "G awareness" check. This is done to ensure the anti-G equipment is functional, and that the pilot is physiologically prepared to pull Gs.

Once this is accomplished and everything checks out okay, we accelerate to 480 knots and descend to 500 feet above ground level (AGL). This speed affords a good compromise between maneuverability and fuel economy. Enroute to the target we practice avoiding. It's low tech and always works for defeating threats. All of the maneuvers are done individually and as a group, the objective always being to get your group through to the target without any losses. As we approach the range, I contact the range officer to receive permission to enter with my four-ship. Once cleared, I begin to maneuver the formation to prepare for the first attack. The attack is a high-altitude 45° dive-bombing run. To accomplish this we must begin our "fly up" from approximately 10 miles away. This calls for a steep climb from 500 feet above ground level to approximately 20,000 feet.

Since the jets are merely specs in the sky at that altitude and distance, the range officer transfers control of the attack to me. During the climb, we maneuver the flight into an echelon position. As I "roll in", I call: "Lobo 11 is in from the west". From that position we "roll in" one right after the other. We drop our bombs within seconds of each other and climb steeply for the safety of greater altitude and clear sky.

After the attack we get back together and egress from the threats. When I'm satisfied that we are finished with our attack and have regained our mutual support, I "terminate" the attack. Any unusual occurrences are quickly discussed on the radio as we reenter the range for the more conventional curvilinear pattern. From this pattern we will deliver the balance of our ordnance from a variety of altitudes and attack angles. As we descend, the range officer resumes control of the range and we continue our individual attacks.

The process is repetitive; it allows more experienced pilots to reacquaint themselves with "sight picture" of each delivery. New pilots strive to learn and memorize what each attack will entail. This process is repeated daily throughout the world, as fighter pilots from all nations hone the skills that their country and Air Force (yes, Navy guys too) hope are never required yet are vitally necessary in these ever changing times.

After the last bomb is dropped we reform as I prepare to lead the formation home. A quick goodbye to the ranger, an appeal to have the bomb scores faxed home (they're needed to verify who won the bet), and we're off to Kansas Center for clearance home.

The trip home is slower paced. There is no excitement, no anticipated events to accomplish. As long as the weather holds together we should be debriefing at home in about an hour. But this being the northern portions of the Mid West, you can never be too relaxed about the weather. Unforecast snowstorms can quickly send us to scurrying for usable airfields. I can easily wind up in Iowa, Nebraska or North Dakota; it wouldn't be the first time or the last.

But the weather gods are looking favorably on us today. We land uneventfully and park our jets on the ramp. Maintenance debrief is our first stop on the way back to Ops. There we fill out our time sheets and discuss any discrepancies we have with the jets.

After that, we drop our flying gear in life-support. The debrief takes place in the same room we briefed the flight. This process can take longer than the flight itself. We carefully critique the flight planning, weaponeering, briefing, and in-flight conduct. Each one of the attacks is carefully reviewed for compliance with parameters and regulations.

After everyone's questions have been answered, it's lunchtime. After all, it's now about 1:00PM. The whole process took about 6 hours. It was hard work, but work I wouldn't trade it for anything. I leave the debriefing with a smile on my face and a couple of extra coins in my pocket.

119th Fighter Wing
North Dakota

119th Fighter Wing, North Dakota (emblem approved 20 March 1968). The bar in the center of the emblem is symbolic of the sky and clouds, which are the elements of aviation. The head of the central arrow is symbolic of a compass needle, while the two crossed arrows represent military security, hence the motto "Guarding the Northlands". The gold and green colors of the background represent the rich Dakota farmlands and the grain produced there.

178th Fighter Squadron, North Dakota (emblem approved 6 December 1949). The golden eagle is native to the state and together with the lightning bolt represents air superiority and the ability to strike quickly and with devastating effects against an enemy or prey. The unit motto is, "Si Vis Pacem, Para Bellum" (If you wish peace, prepare for war). The squadron is known as the "Happy Hooligans",, a name which is painted on the tails of their aircraft.

178th Fighter Squadron, North Dakota "Happy Hooligans".

This name derived from one of the early unit commanders "Pappy" Larson who chided the pilots for their aerobatics, cowboy antics, and other flying skills learned in World War II as being a "bunch of hooligans". The name remained as a favorite among the pilots and is still the unofficial motto of the unit.

Constituted at Hamilton Field, California in May 1943, activated as the 392th Fighter Squadron, assigned the 367th Fighter Group, Ninth Air Force, the squadron flew Lockheed P-38J Lightnings in the ETO beginning in May 1944 and then in Republic P-47D Thunderbolts from February 1945 to VE Day. The squadron was reconstituted and redesignated as the 178th Fighter Squadron and allotted to the North Dakota ANG on 24 May 1946.

On 16 January 1947 the squadron was equipped with P-51 D Mustangs at Hector Field, Fargo municipal airport. The unit was mobilized for the Korean War on 1 April 1951 and moved to George AFB, California.

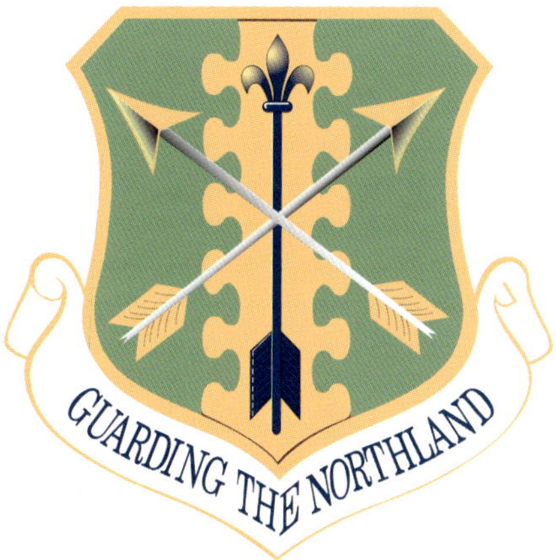

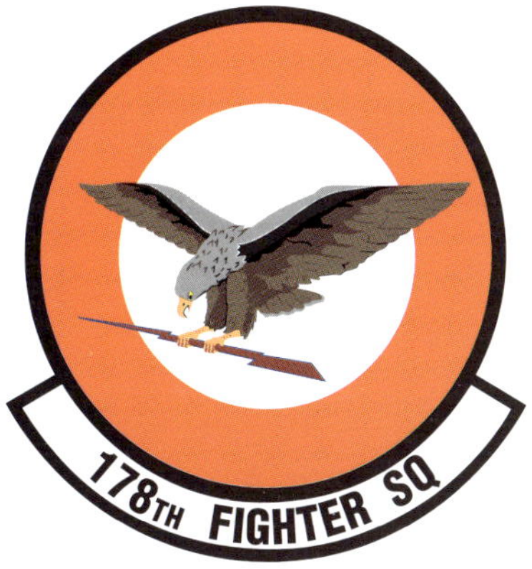

Returned to state control on 1 December 1952, during the Summer 1954 the squadron converted to F-94A/B Starfires. On 16 April 1956, the 119th Fighter Group was established as the unit's headquarters.

From the Summer 1958, the squadron converted to Northrop F-89D Scorpions, then to Convair F-102 Delta Daggers in the Fall 1966, in the Fall 1969 to McDonnell F 101 B/F Voodoos. From Spring 1977 to 1978 the unit converted to McDonnell F-4D Phantoms. They became Tactical Air Command gained on 1 April 1980. From April 1986 to April 1987 the squadron provided aircraft and personnel for Operation *Creek Klaxton*, the ANGs assumption of the air defense mission to US Air Force Europe (USAFE).

They provided air cover for the regular Air Force while the 86th Tactical Fighter Wing, Ramstein AFB, Germany, converted to F-16C/D Fighting Falcons. In 1990 the squadron converted onto F-16A/Bs as Air Defense Fighters with a forward detachment located at Klamath Falls, Oregon.

On the right
Lt.Col. Tom Larson with Capt. Rick Omang check a 119th Fighter Wing F-16 ADF.

In the page on the right
Capt. Rick Omang eases himself in his Viper cockpit.

Further on the right
The exaust vanes of a Pratt & Whitney F110 augmented turbofan.

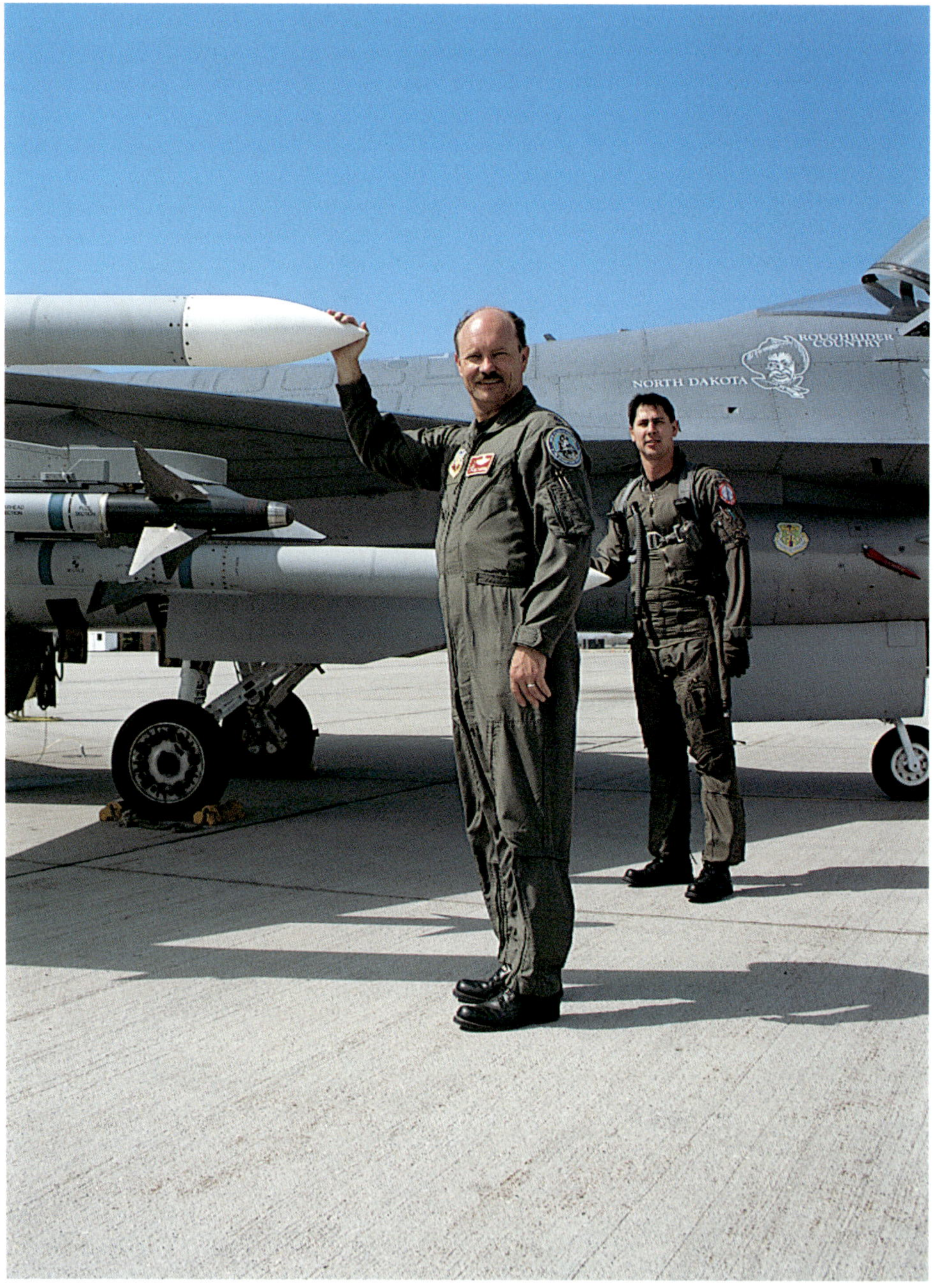

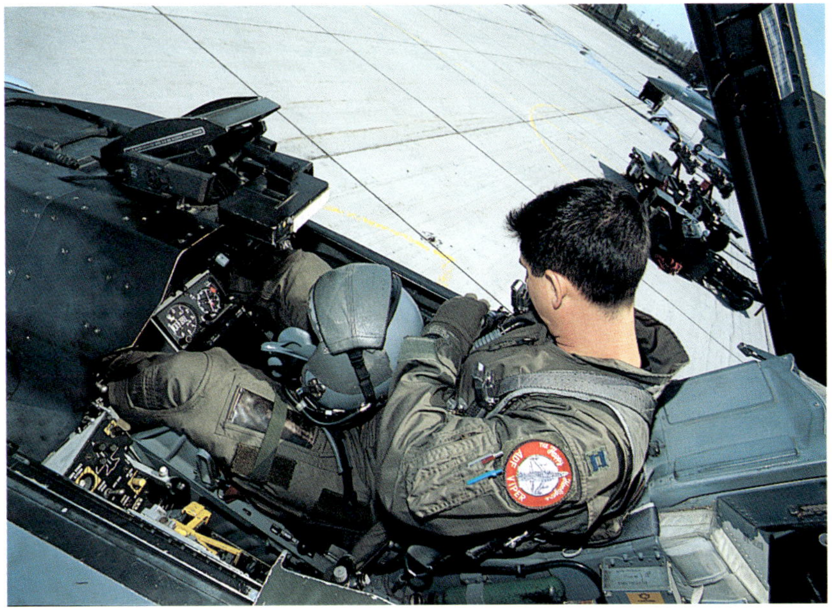
ADF
VIPER

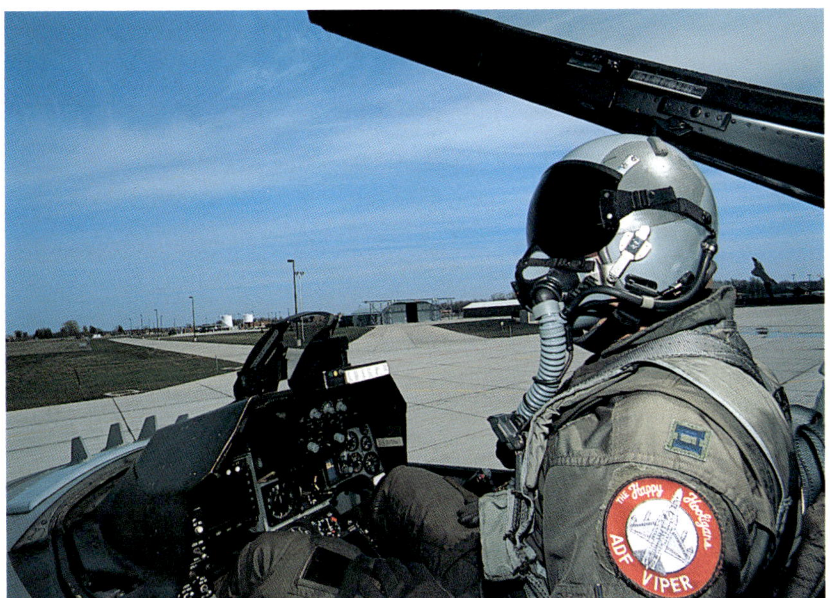
The Happy Hooligans
ADF VIPER

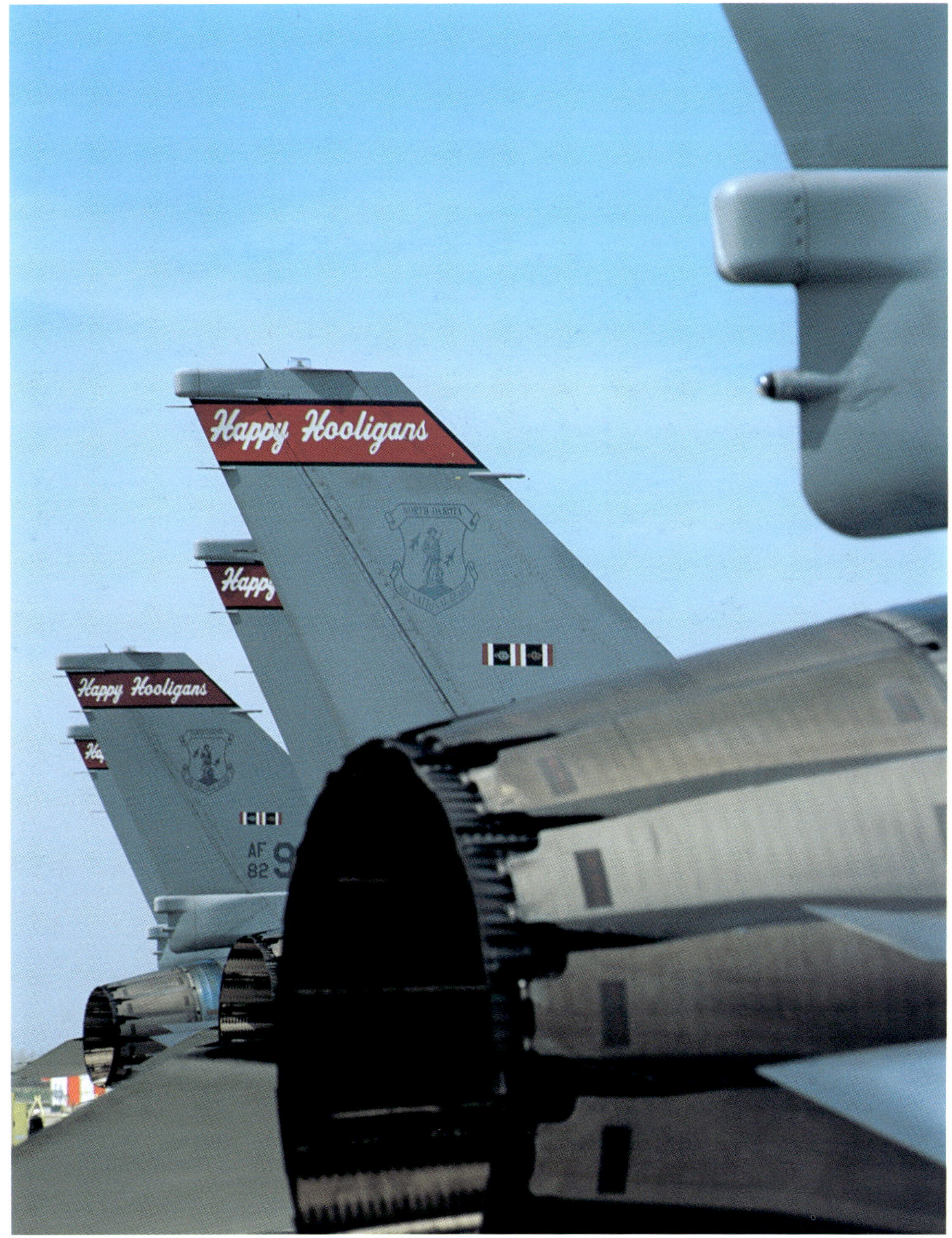
Happy Hooligans
Happy Hooligans
AF
82

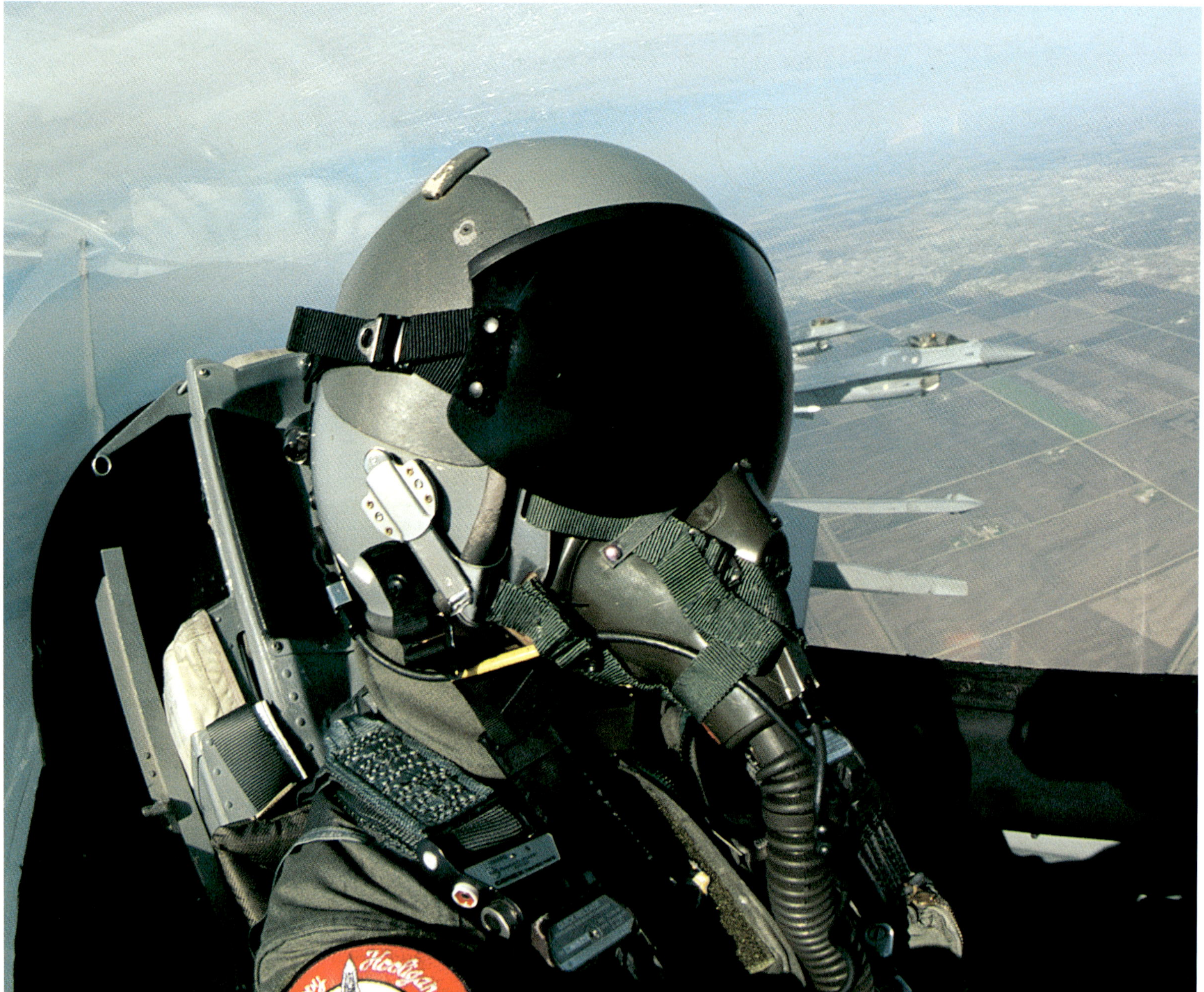

Above A 119th Fighter Wing pilot in flight vith two other Vipers.

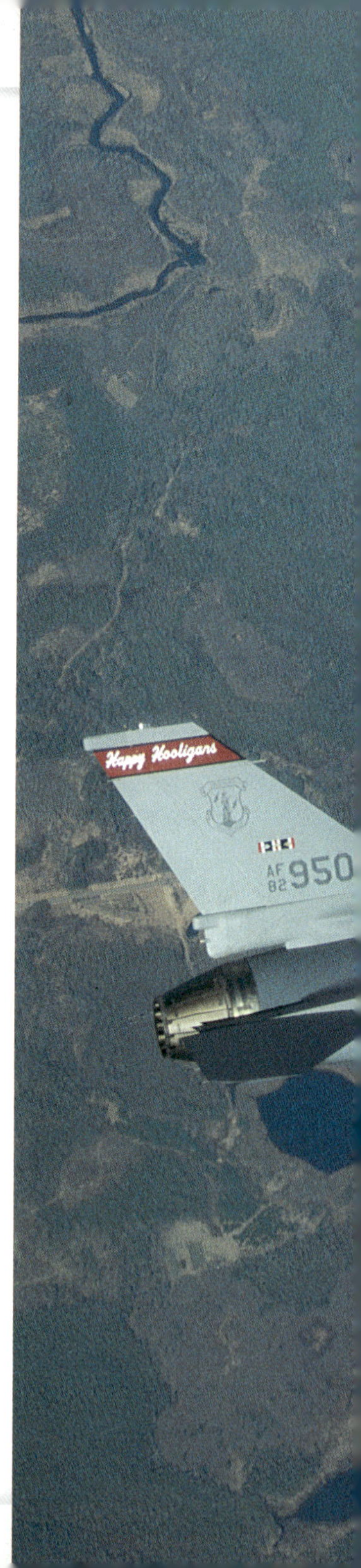

On the right Two 119th FW F 16s turn over some Minnesota ponds.

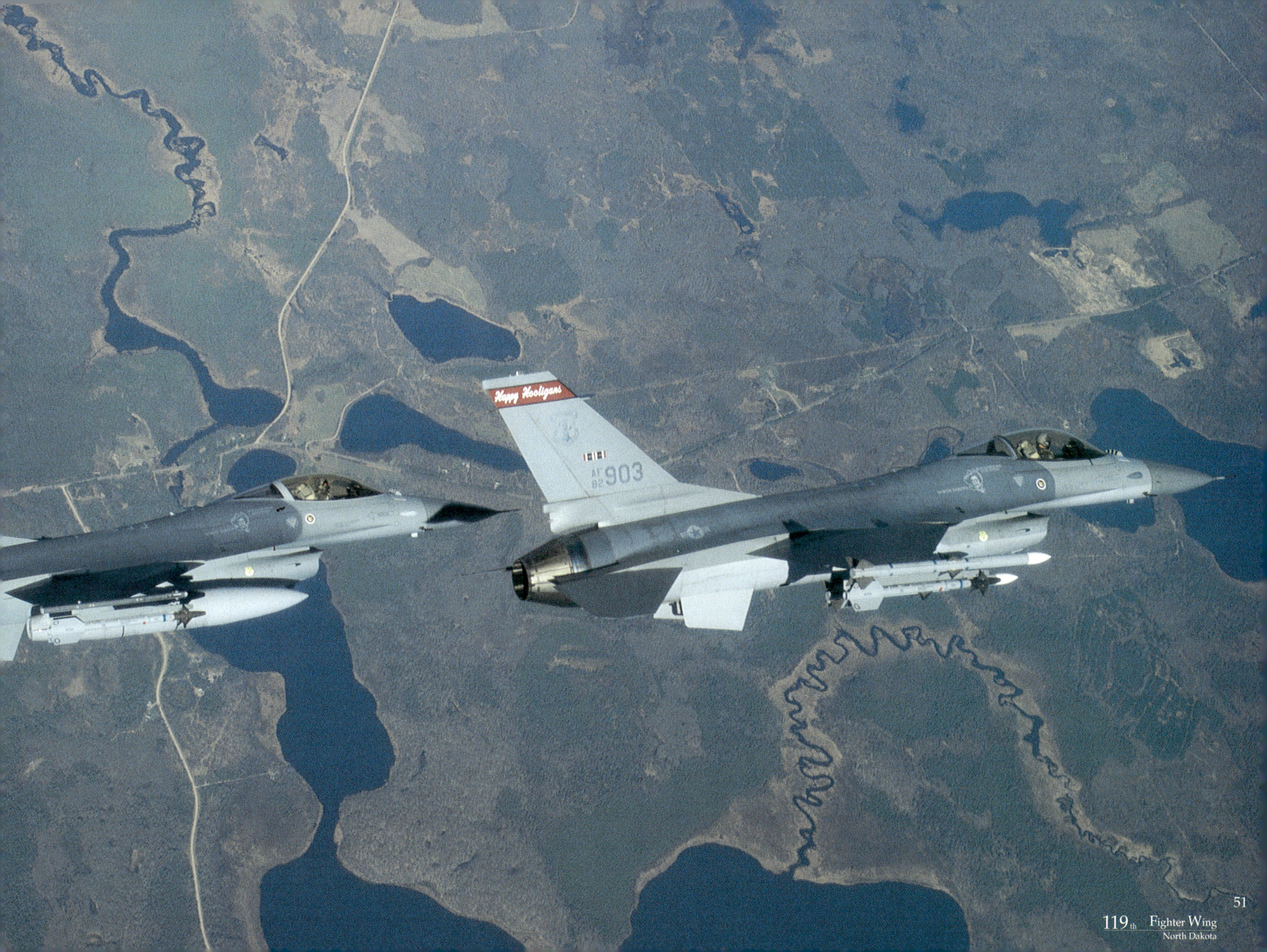
Happy Hooligans
AF 82 903

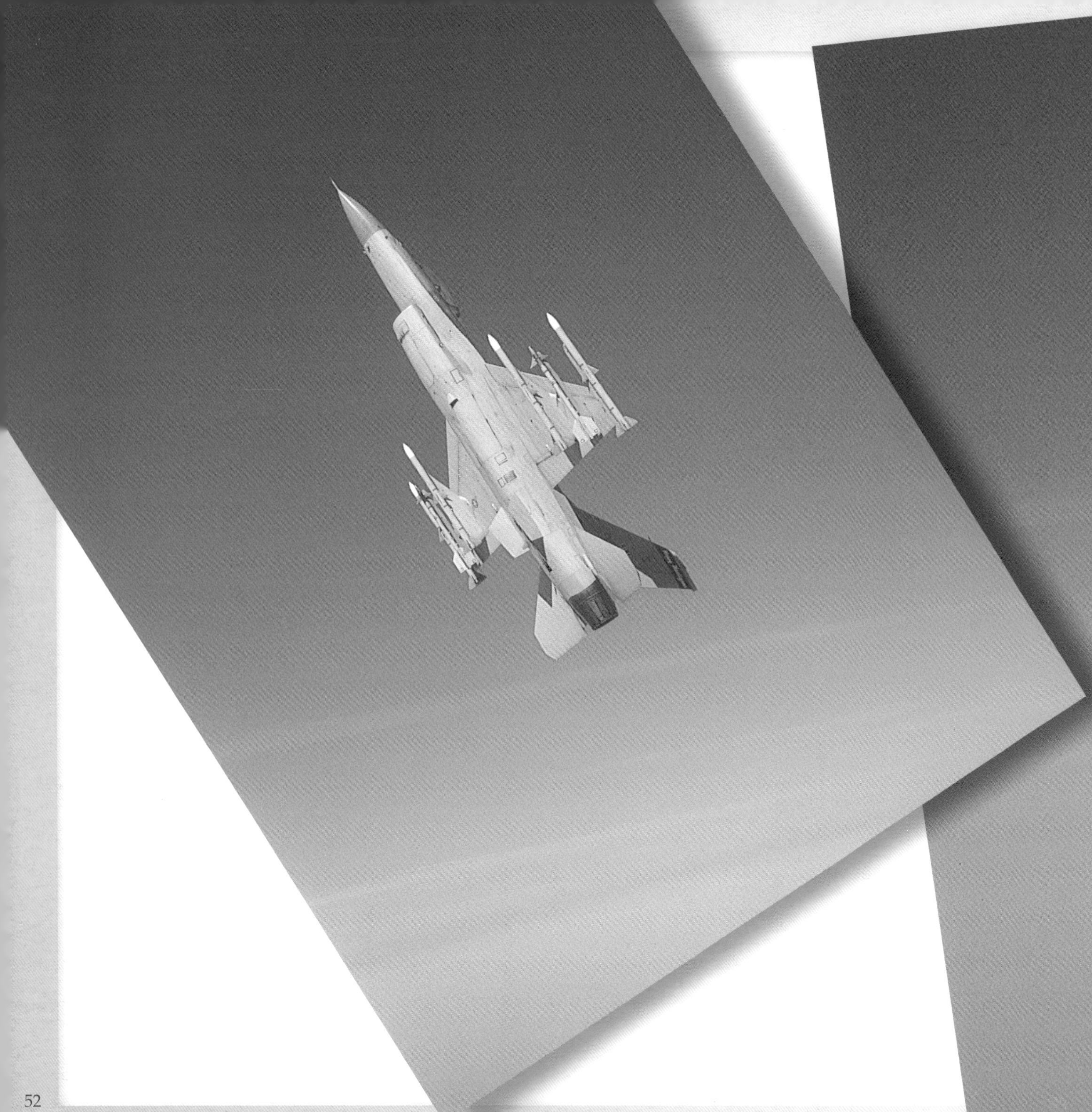

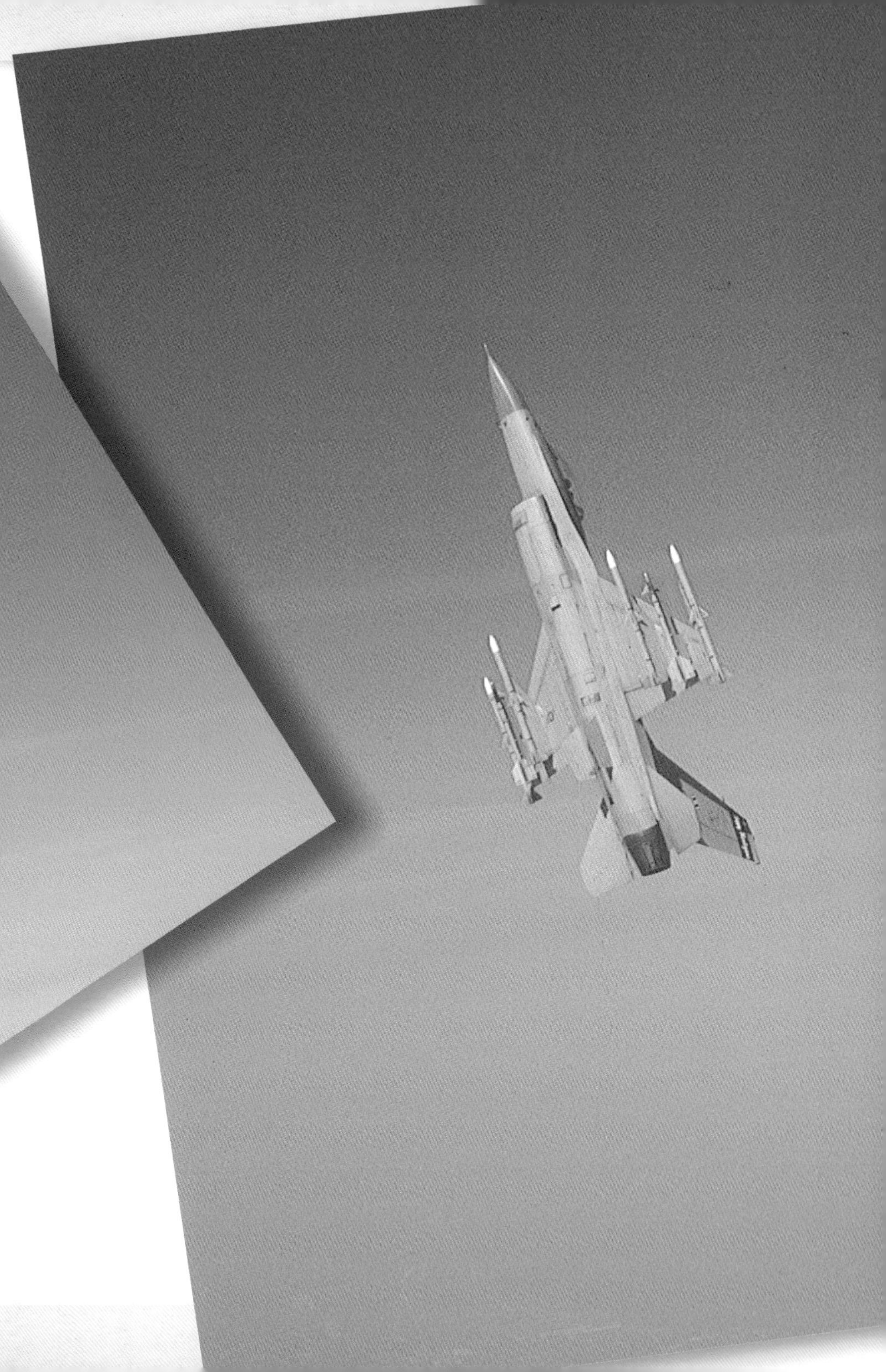

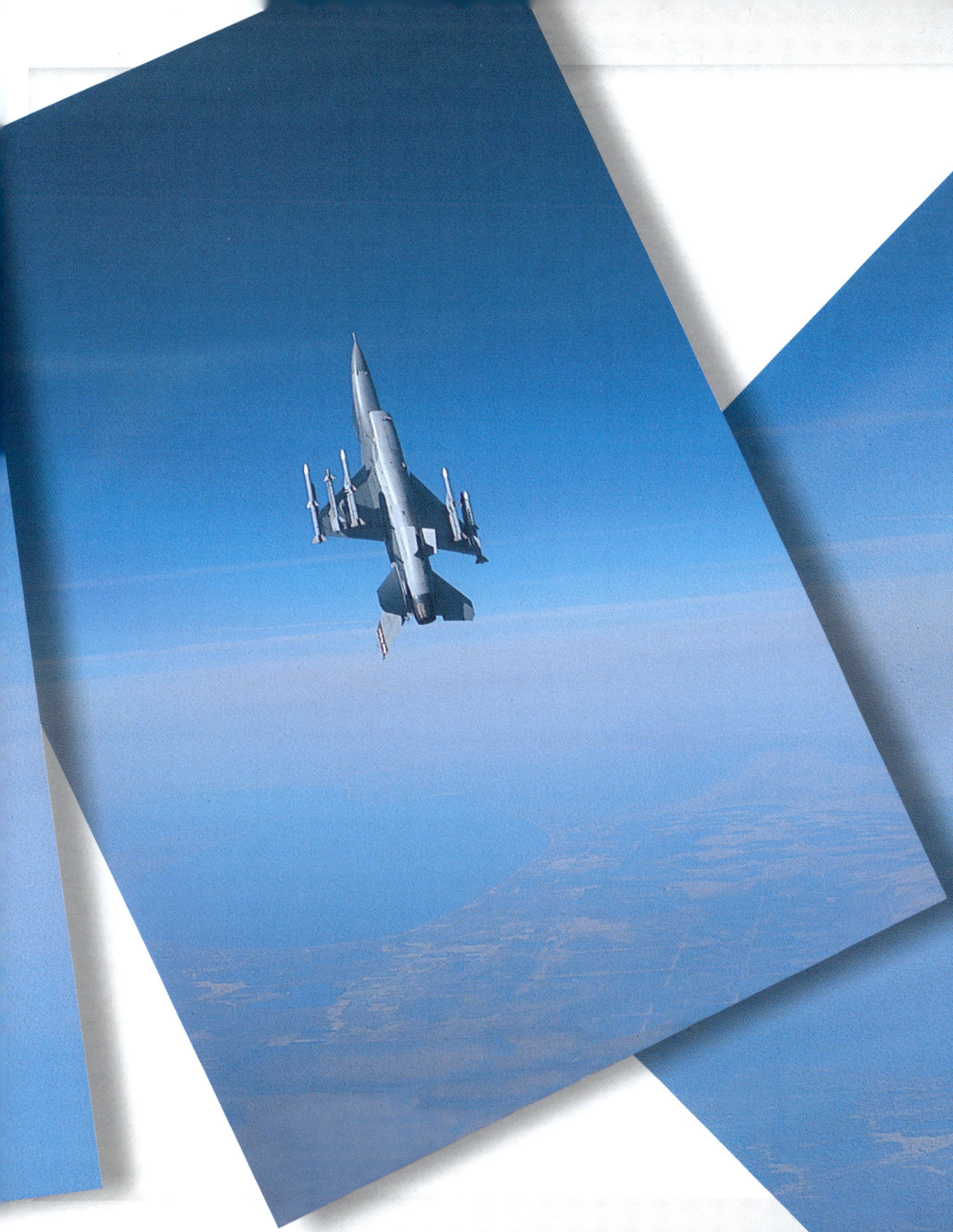

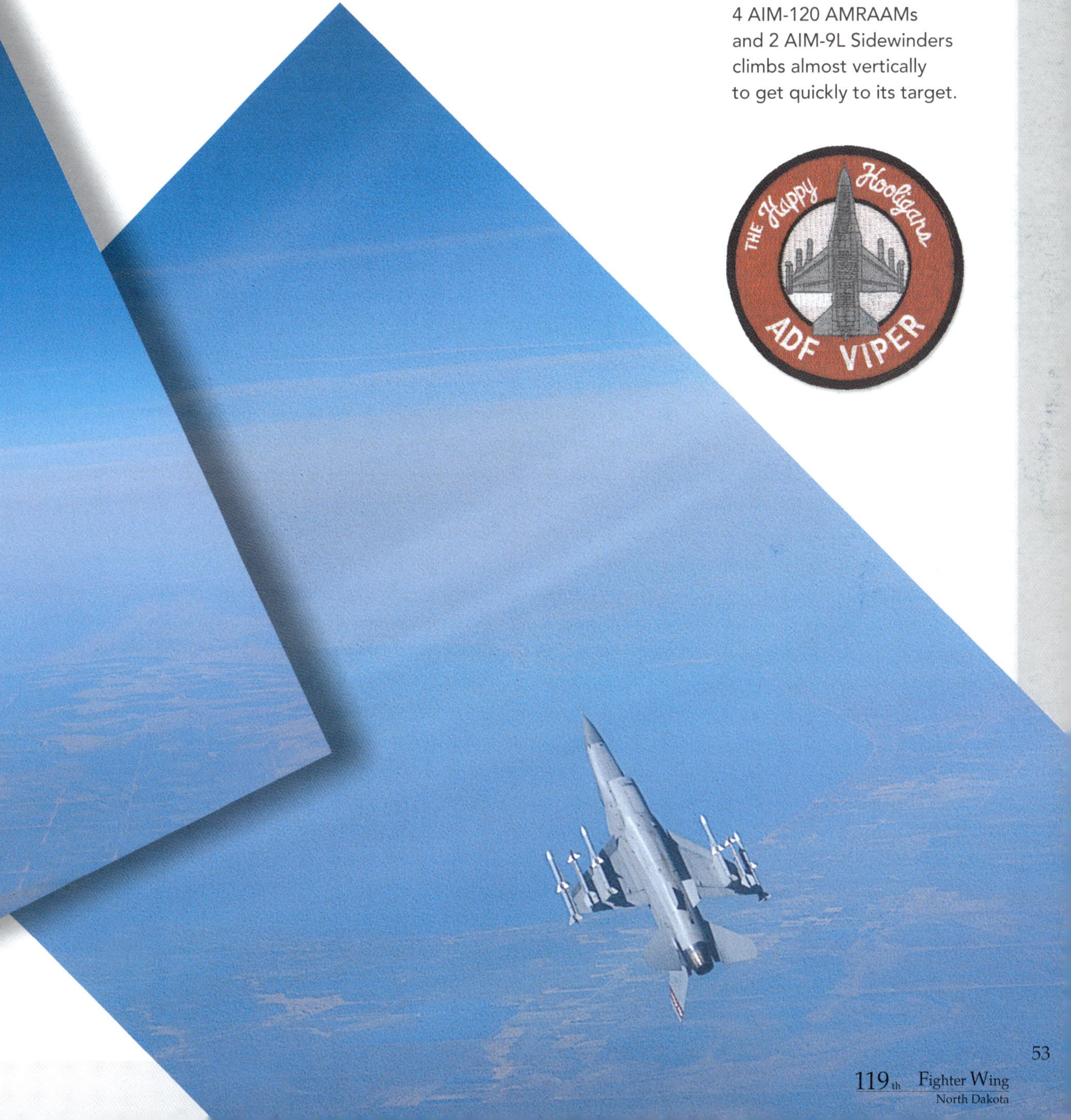

A "Happy Hoolingans" F-16 ADF armed with 4 AIM-120 AMRAAMs and 2 AIM-9L Sidewinders climbs almost vertically to get quickly to its target.

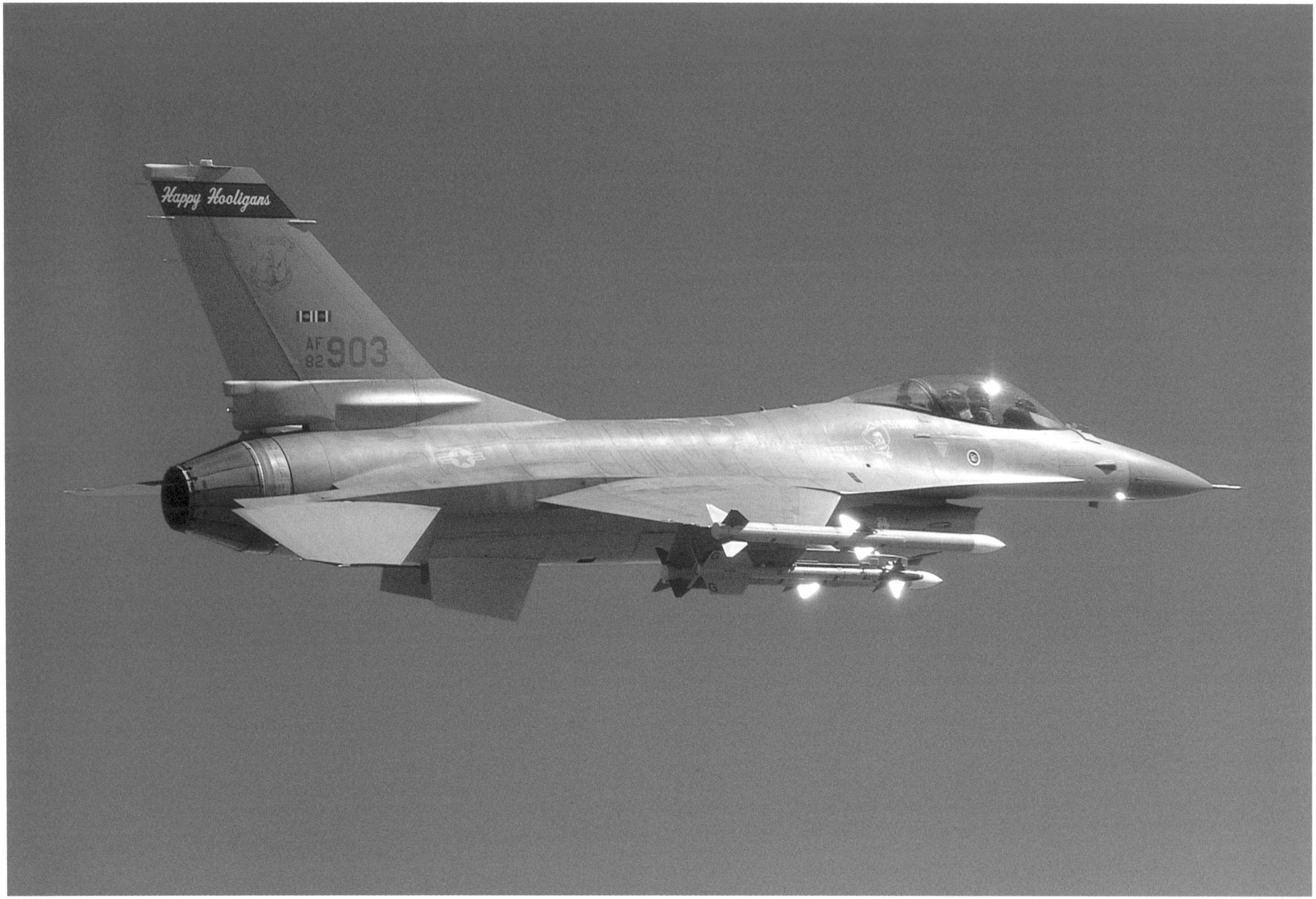

Abpve A 119th FW Viper in air-intercept configuration is lit-up by the high-altitude sun.

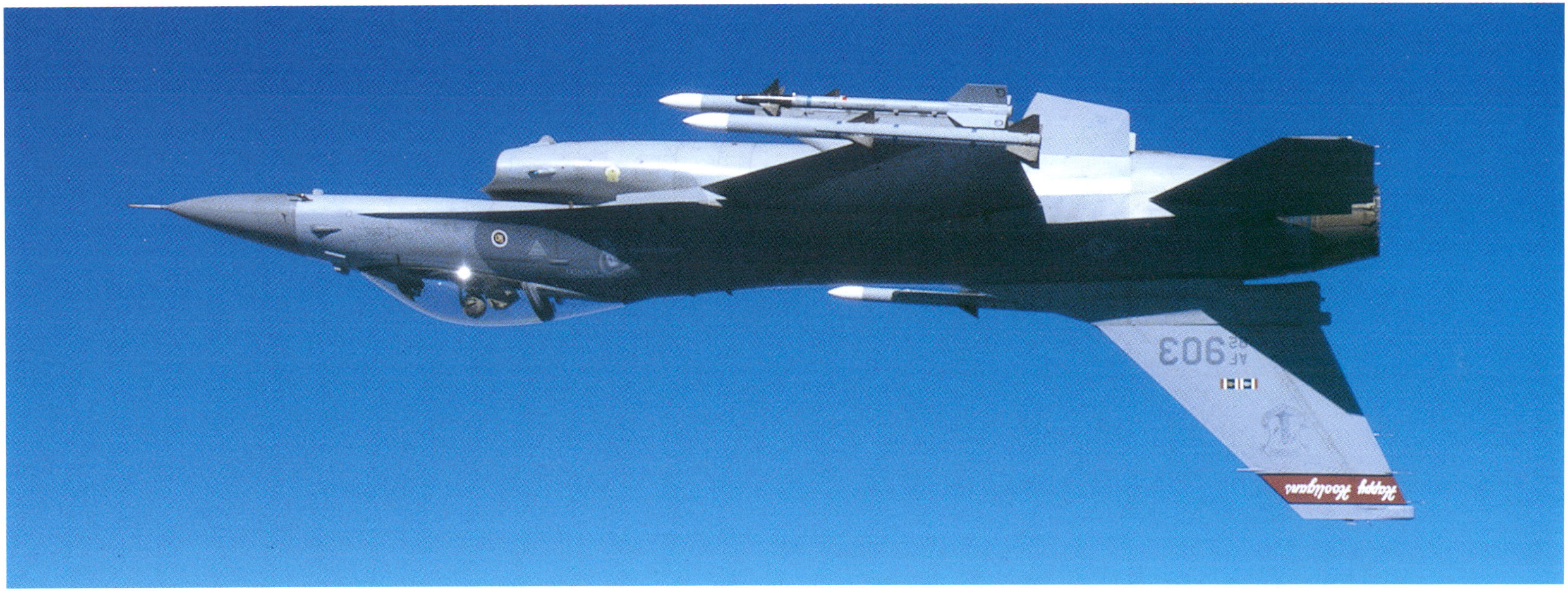

Above Upside down on top of a loop.

"HAPPY HOOLIGANS" ON DUTY

BY DAVID SOMDAHL

It began with a call for help. At the end, it marked a day several Hooligan pilots will not forget.

Pilots from the 178th Fighter Squadron, 119th Fighter Wing, ended up locating and escorting a civilian business jet that deviated from its flight plan. The Learjet 35 jet crashed near Mina, S. D., carrying world-renowned professional golfer Payne Stewart and five others.

The jet took off from Orlando, Florida., that morning bound for Dallas, Texas. Somewhere above Gainesville, Florida., it began flying a straight course to the northwest. Instead of arcing toward Dallas, it barreled on without turning until crashing in a field in South Dakota.

Maj. Kent Olson and Capt. Rick Omang were making preparations for a mid-afternoon training sortie. Lt. Col. Thomas Larson, the operations group commander, was attending to last minute business before preparing for a cross-country flight to Tyndall AFB, Florida.

The business with the Learjet in distress disrupted those plans. "We had one jet available and the morning 'go' was just coming back for hot pit refueling," Olson said. One jet from the morning flight was refueled, then shut down for Omang to set up for an alert launch.

Olson followed Omang out, bringing map kits to the two aircraft. "When we walked back in, we were told, 'You're on. Hit the bricks,'" Olson said.

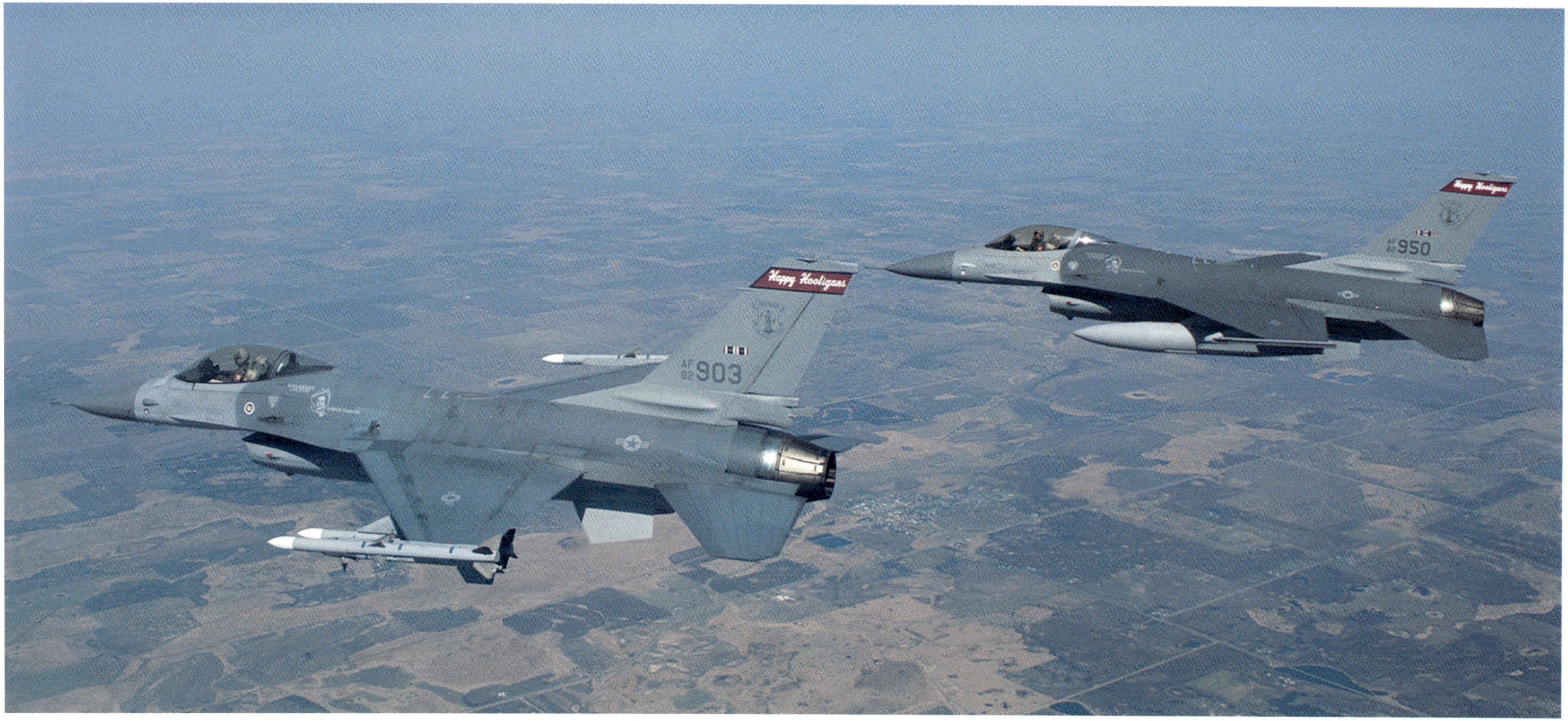

"So out of the door we went. It was that quick." Initially Omang and Olson were told the aircraft was flying about 41,000 feet and was unresponsive to radio calls placed by Federal Aviation Administration controllers. "We knew it was a Learjet and that was about it," Olson said.

Prior to flying full-time for the Hooligans, Olson flew Learjet 35 and 36 jets for two and a half years.

Phoenix Air, a defense contractor, provided on-demand aerial targets and airborne electronic countermeasures systems, flying against Air National Guard, Air Force and Navy pilots in the Faker Program. Drawing on his own personal flight experience Olson said, "I knew they would likely run out of fuel somewhere over North Dakota or South Dakota."

Olson, an amateur radio operator, fired up a high frequency radio transmitter-receiver installed on all F-16 air defense fighter aircraft. He was able to talk with military controllers at the Western Air Defense Sector, McChord AFB, Washington, while still on the ground prior to take-off.

Omang checked in with FAA controllers from Minneapolis Center on another

In the page on the left
Two Vipers in formation for an air defense intercept mission.

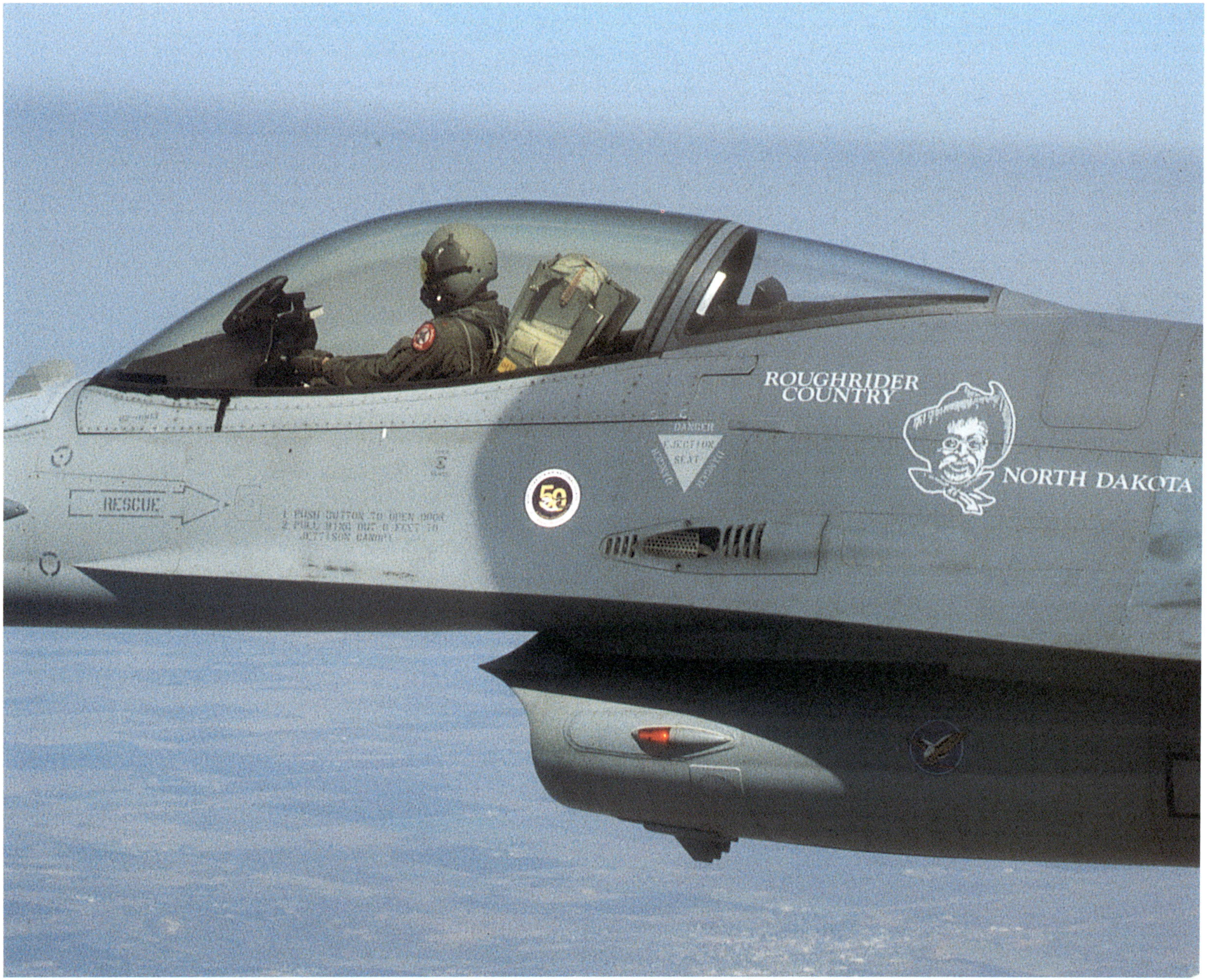

On the right
Combat air patrol missions call for heavy pilot workload and are likely to produce a lot of stress.

radio, and the two compared information as they streaked south, away from Fargo. As they approached the Iowa - Missouri state line, Omang and Olson made contact with the Lear.

Instead of finding a single aircraft, they also found a tanker from the 190th Air Refueling Wing, Kansas Air National Guard, refueling a pair of F-16C fighters from the 138th Fighter Wing, Oklahoma Air National Guard.

The Lear was now flying nearly 49,000 feet high. During this time a second pair of F-16s launched from Fargo on the escort mission. The second flight, piloted by Maj. Craig Schroeder and Capt. William Cowden, was meant to relieve Olson and Omang.

Olson talked about the difficulty of getting close to the jet, noting "we've never practiced an intercept of this type at that altitude." Olson and Omang lit the afterburners on their jets, but only Olson's jet had sufficient thrust to maneuver for a close look. "I actually got out in front of the target and about 500 feet below him," Olson explained.

Olson made "S-turns" to bleed off energy in an effort to maneuver alongside. Instead, he lost altitude and pulled away to the left. The pair continued to radio back altitude, airspeed and bearing to FAA controllers and to "Bigfoot" controllers at McChord.

A few minutes later Omang broadcast, "Hey, we've got some movement." Omang was further behind and lower when it appeared the Lear ran out of fuel. "I could tell he had flamed out by the way the plane reacted," Omang said. "One engine flamed out first, then it started to turn off to the right." Seconds later the left engine quit, then it began doing aileron rolls as it fell nearly straight down toward the earth.

"He was going down pretty fast," Omang added. "When he did go out of control, it was pretty eerie." Reviewing cockpit videotapes after the flight, Omang and Olson guessed that the Learjet's rate of descent was greater than 30,000 feet per minute.

The plane impacted the ground about 600 miles per hour, according to a statement made public by the National Transportation Safety Board. The Fargo pilots were able to watch the plane for about a minute, losing sight of the Learjet as it fell past 20,000 feet.

After being cleared by air traffic controllers, Olson and Omang immediately began searching for the crash site. They found a small, darkened spot in a farm field - there was a minimal amount of smoke, Olson said, similar to someone burning leaves in a yard.

The pair radioed back the longitude and latitude, staying overhead until the second flight from Fargo arrived. They witnessed the arrival of two people in a truck, possibly hunters from South Dakota's pheasant hunting season. A short time later, firefighters and a medical evacuation helicopter arrived at the site.

Omang said he thought they would be able to help a misdirected pilot, someone who was lost. "By the time we got up there, it was pretty apparent that was not necessarily going to be our mission," Omang said. " We were there to monitor and watch him."

138th Fighter Wing
Oklahoma

138th Fighter Wing, Oklahoma (emblem approved 28 August 1967). The Air Force blue and yellow with flight symbol dividing the shield represent the unit's day and night operational ability. The globe represents the unit's world-wide flying operational ability, while the Indian shield represents the State of Oklahoma. The green drops on the Indian shield represents the unit's location in Tulsa, the "oil capital of the world". The unit motto is "Strength through Preparedness".

125th Fighter Squadron, Oklahoma (emblem approved 15 September 1943). The "eager" beaver's expression portrays the happy and confident attitude of the squadron's personnel. The beaver's attire represents the outstanding personal appearance of the squadron's personnel. The three cards indicate that the squadron is playing a "top" hand for a winning game.

125th Fighter Sqn., Oklahoma "Eager Beavers". Designated the 125th Observation Squadron was allotted to the State of Oklahoma on 30 July 1940. Located at Tulsa, the squadron had two Douglas O-38Es, one North American O-47B and one North American BC-1A observation aircraft. Assigned to the 68th Observation Group, the squadron moved to Post Field, Fort Sill, Oklahoma on 31 November 1941. The unit remained in the continental United States (CONUS) until May 1944 when it was shipped to England. It was then equipped with Stinson L-5s and flew combat operations in support of ground forces in Northern Europe from August 1944 until VE Day. The unit was reconstituted and redesignated as the 125th Fighter Squadron and allotted to the Oklahoma ANG on 24 May 1946. The squadron was equipped with North American P-51D Mustangs at Tulsa municipal airport on 15 February 1947.

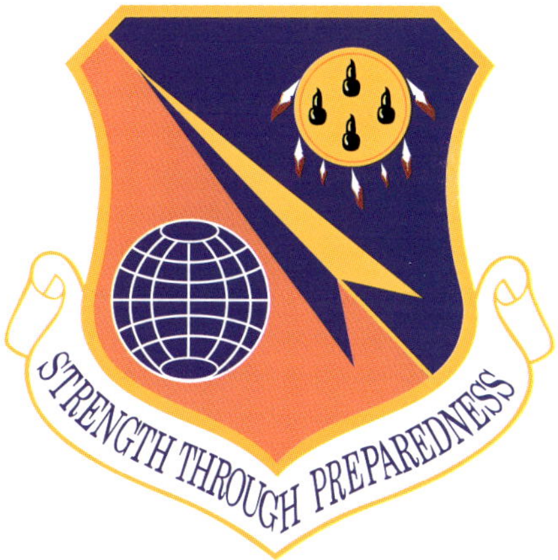

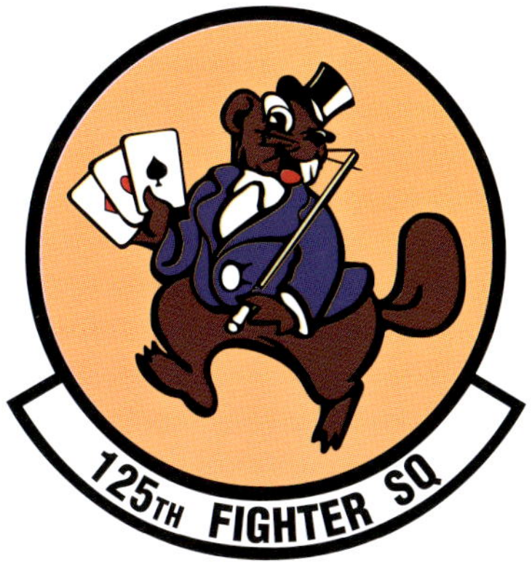

The unit converted to Republic F-84B Thunderjets in June 1950. Mobilized for the Korean War on 10 October 1950, it moved to Alexandria AFB, Louisiana in November 1950. Assigned to the 137th Fighter-Bomber Group, it returned to state control on 10 July 1952 when it was converted back to F-51D Mustangs.

In 1954 it converted to Lockheed F-80C Shooting Stars. On 1 August 1957 the 138th Fighter Interceptor Group was established. During the same year the unit converted to North American F-86L Sabre Jets. From Winter 1959 to Spring 1960 the unit converted from Sabre Jets to four-prop Boeing C-97G Stratofreighters. On 15 January 1960 the unit was redesignated the 125th Air Transport Squadron and was gained by Military Air Transport Service (MATS) on 1 July. The squadron was mobilized, on 1 October 1961 for the Berlin Crisis to conduct worldwide airlift missions from its base in Tulsa. Returned to state control on 31 August 1962, the squadron converted to Douglas C-124C Globemaster IIs. In 1973 the unit converted to North American F-100D/F Super Sabre and back to Tactical Air Command. After converting to A-7D Corsair IIs in July 1978, the unit was re-equipped with F-16C/Ds in 1993 under Air Combat Command in June 1992.

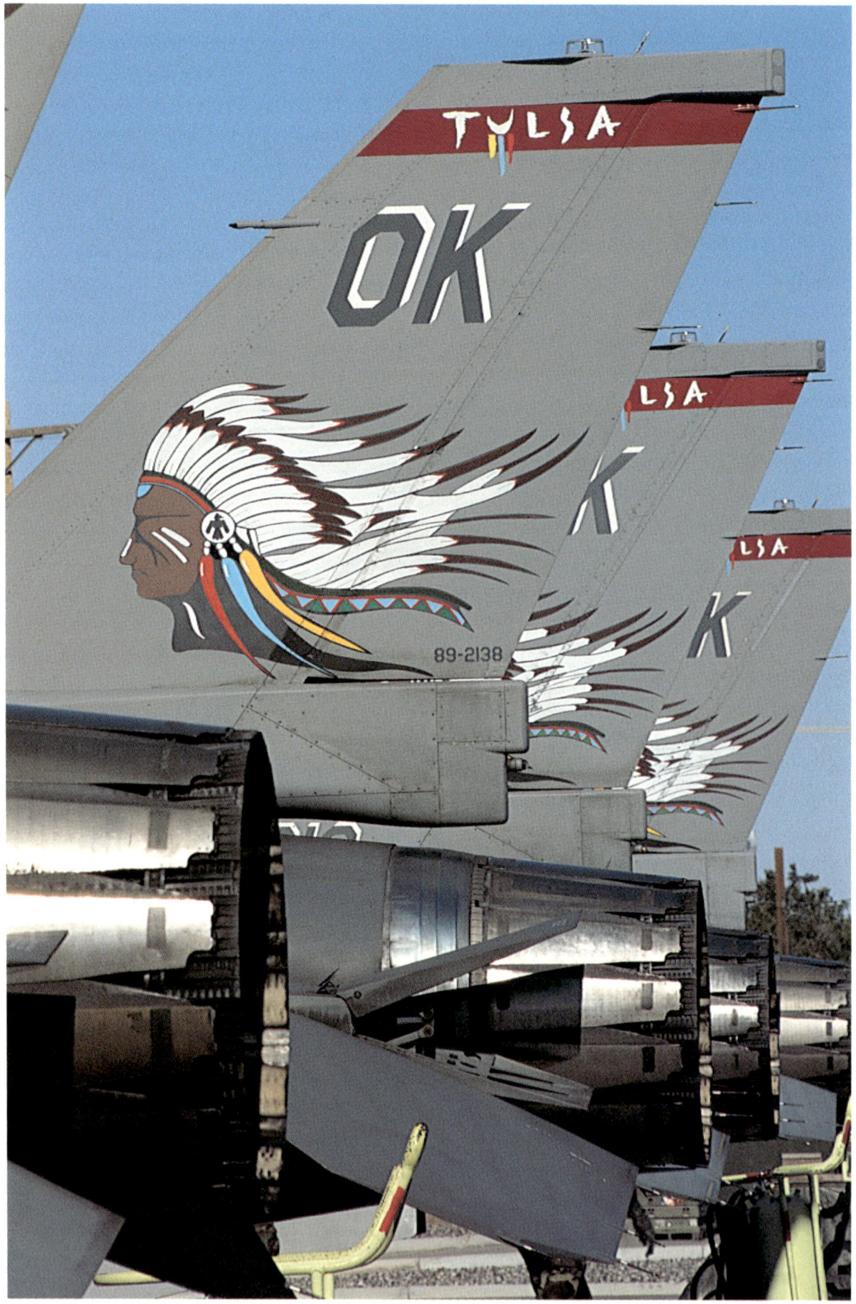

NORTHERN WATCH

BY CAPT. MARK SCORSONE

The Operation *Northern Watch* package members thought it might be a slow day since it was an Iraqi holiday. Even so, there was a lot of anticipation since just the day before the 138 Fighter Wing had dropped it's first bombs in combat.

Six aircraft from the Tulsa Air National Guard were tasked to be part of the Operation *Northern Watch* package of aircraft. Overall, the package included American and British fighter and support aircraft. The overall mission commander for the day was a Tulsa Air National Guard pilot, callsign "Pigg".

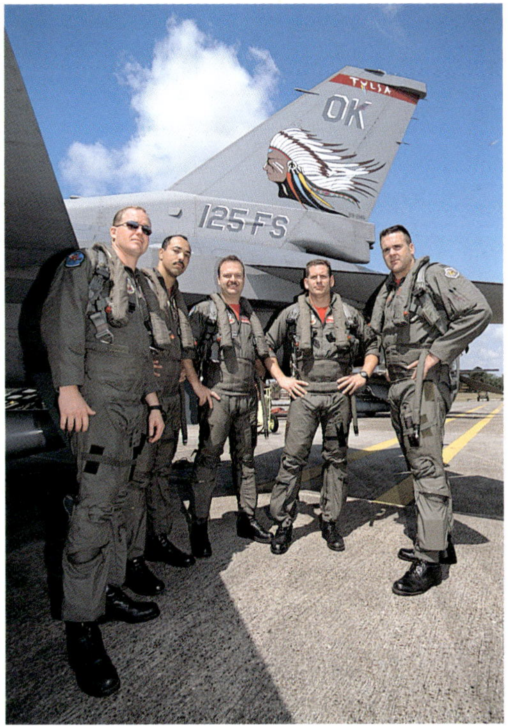

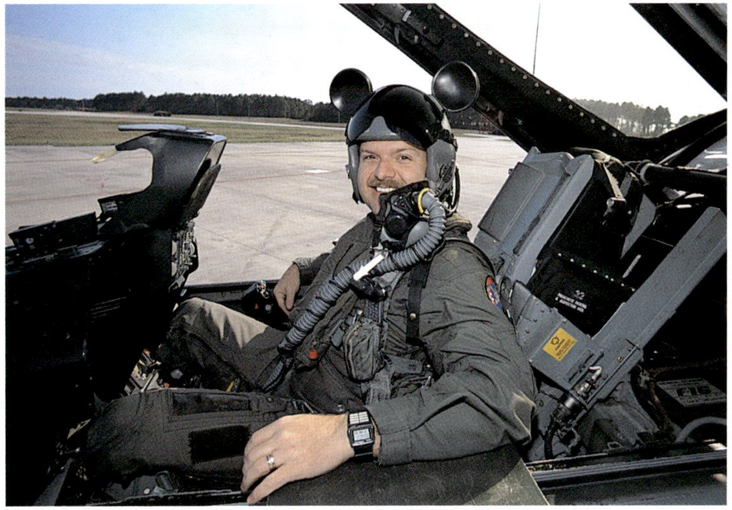

On the left
Capt. Jim Crowley arrives at Tyndall AFB for an air-to-air training week. In the occasion, he added Mickey Mouse ears to his helmet.

Below
Perfectly lined-up on the overcrowded Nellis AFB tarmac, the 138th FW F-16s wait for their turn in the daily "Red Flag" round of sorties.

Above
A group of 125th Fighter Squadron pilots.

On the left
Fins of 138th FW F-16Cs parked at Nellis AFB for a "Red Flag" exercise.

On the right
Capt. Scott "Francis" Key and Dave "Cheese" Burgy mime a dogfight with 325th FW F-15Cs.

In the page overleaf
125th FS Vipers in flight over the Gulf of Mexico.

In the facing page
Three flares are lunched almost simulteneously to ensure cover against IR-guided missiles.

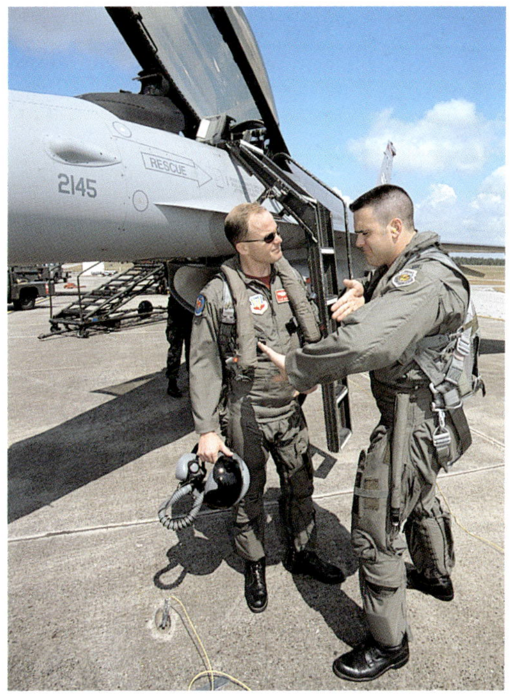

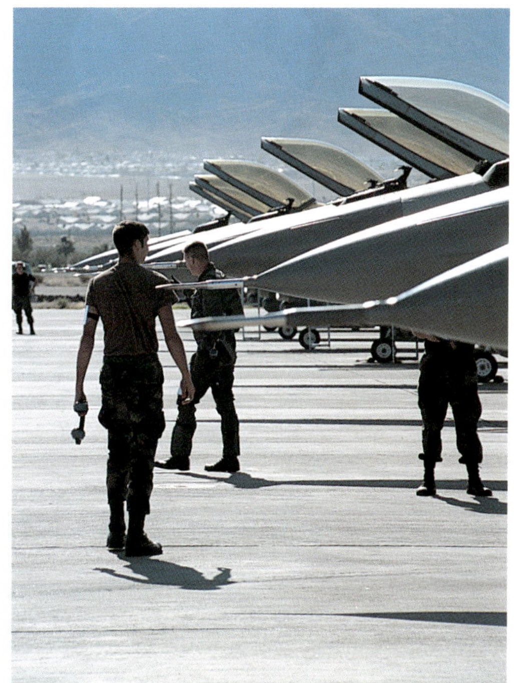

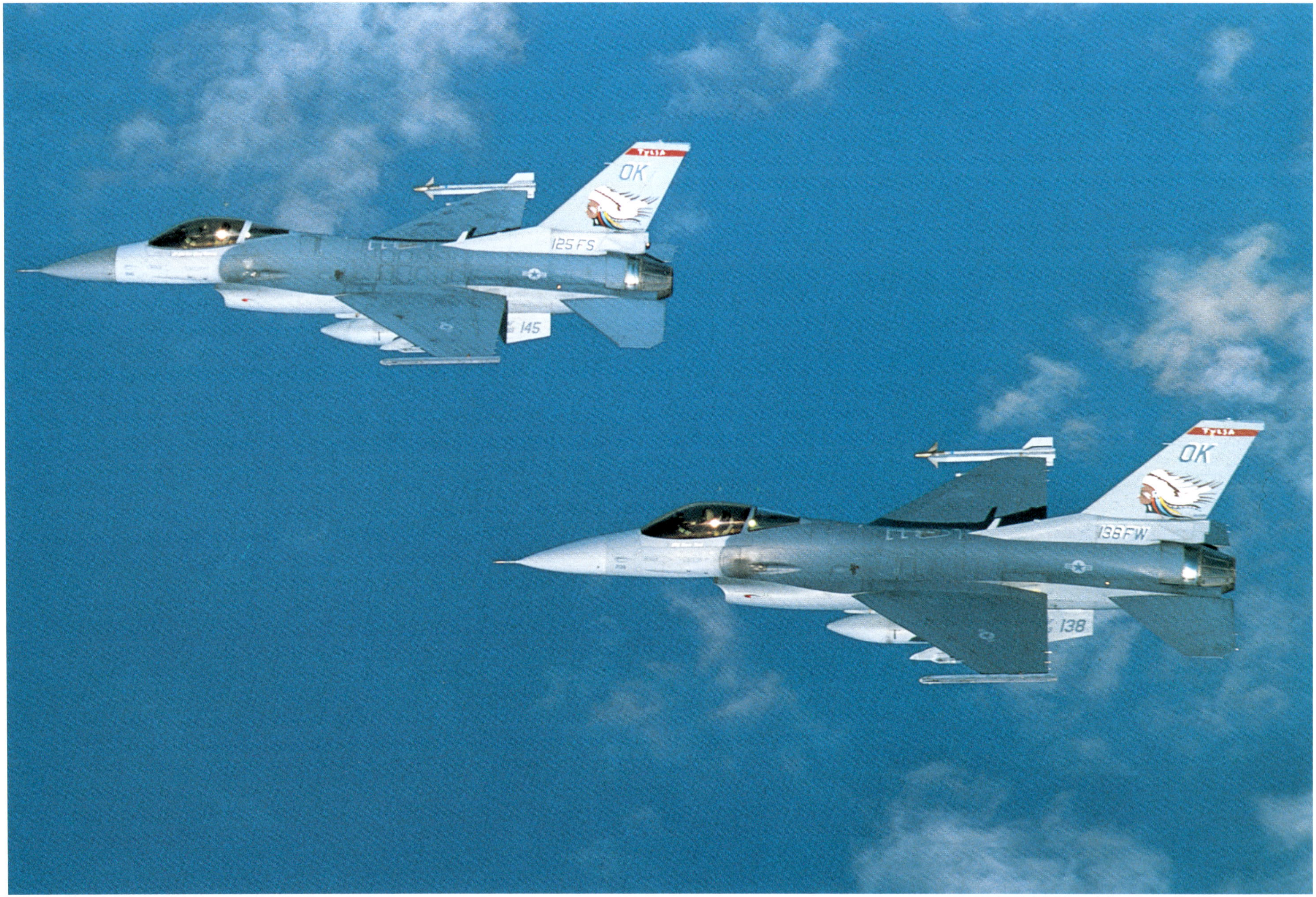
OK
125 FS
145
OK
138 FW
138

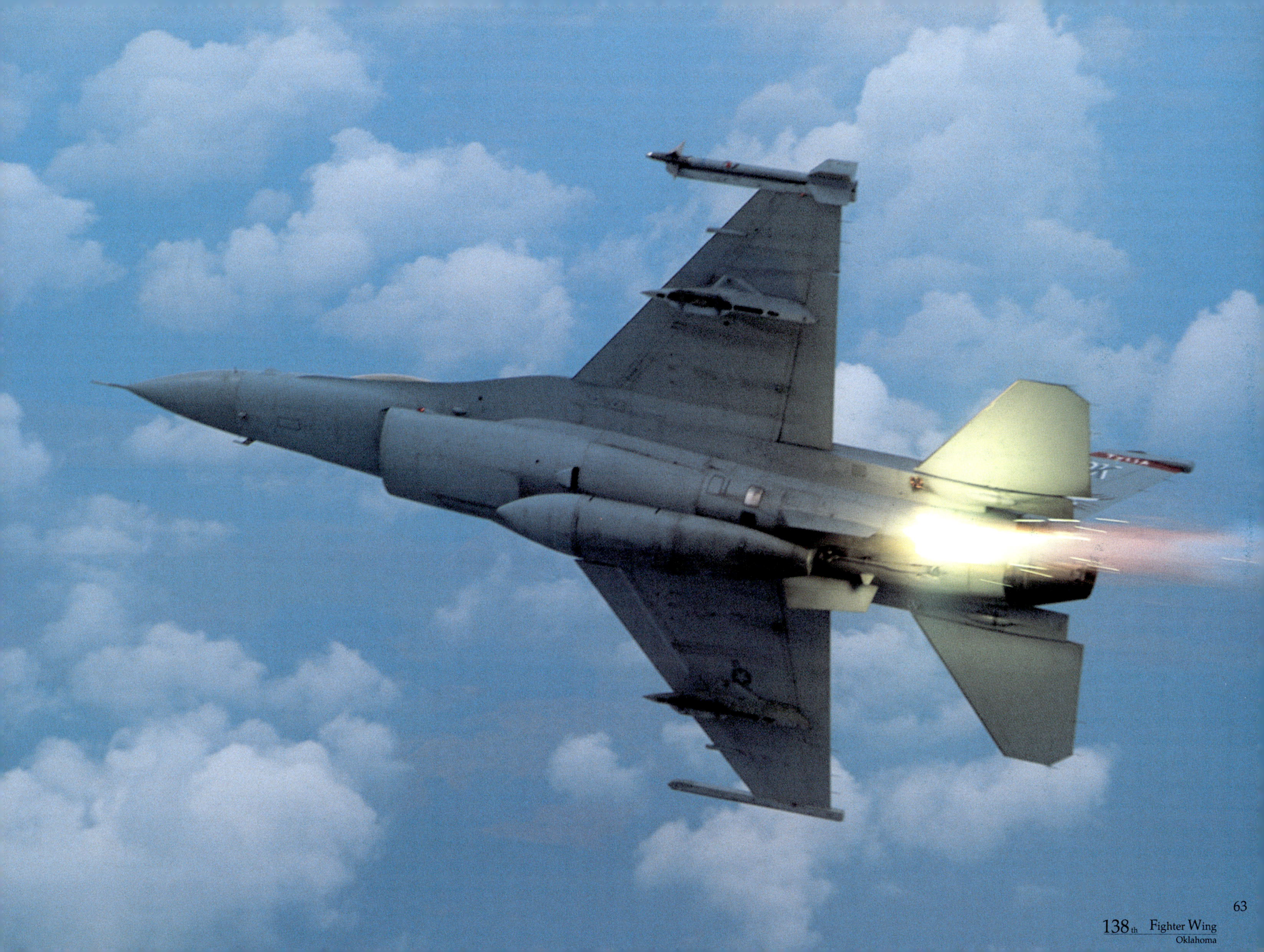

The Tulsa aircraft combined with the F-15E's were tasked as the Emergency Defense Suppression aircraft. If the Iraqi forces demonstrated hostile acts on any coalition aircraft, we were directed to destroy a communications site. This target was a military compound just outside of Mosul, Iraq.

The targets had been positively identified military communication vehicles. Seven targets were briefed and all were communications vehicles.

Pigg assigned specific targets to coalition aircraft and briefed the day's game plan. Shortly after Operation *Northern Watch* aircraft entered the Northern No-fly Zone, the Iraqis fired Anti-Aircraft Artillery (AAA) at coalition aircraft. The first pilot to attack was "Rookie". Rookie released two 500-pound laser-guided bombs and guided them to a direct hit on his briefed target.

The explosion destroyed his target and another vehicle that was parked nearby. The next two-ship was Pigg and "Swamp". Both aircraft scored direct hits on their briefed targets. "Stump" dropped his bombs on the remaining target with another direct hit. Tulsa was now 4 for 4.

With all the briefed targets destroyed, the only 138 Fighter Wing aircraft with bombs remaining were piloted by "Boston" and "Joker". In the midst of air refueling during the attack, the two-ship received a short notice tasking including more targets. A quick review showed two of those targets already destroyed due to the close proximity to previous targets. Again the Iraqi's fired AAA at coalition fighters.

The remaining targets were assigned and Boston's two-ship called attacking from the west. Breaking out the targets through the smoke from the other burning vehicles, Boston and Joker guided their bombs to direct hits.

With all of the briefed and retasked targets destroyed, the 138 Fighter Wing celebrated the Iraqi holiday as one of the finest day's in the unit's history.

On the right Three 125th Fighter Squadron Vipers enroute to Eglin AFB, Florida.

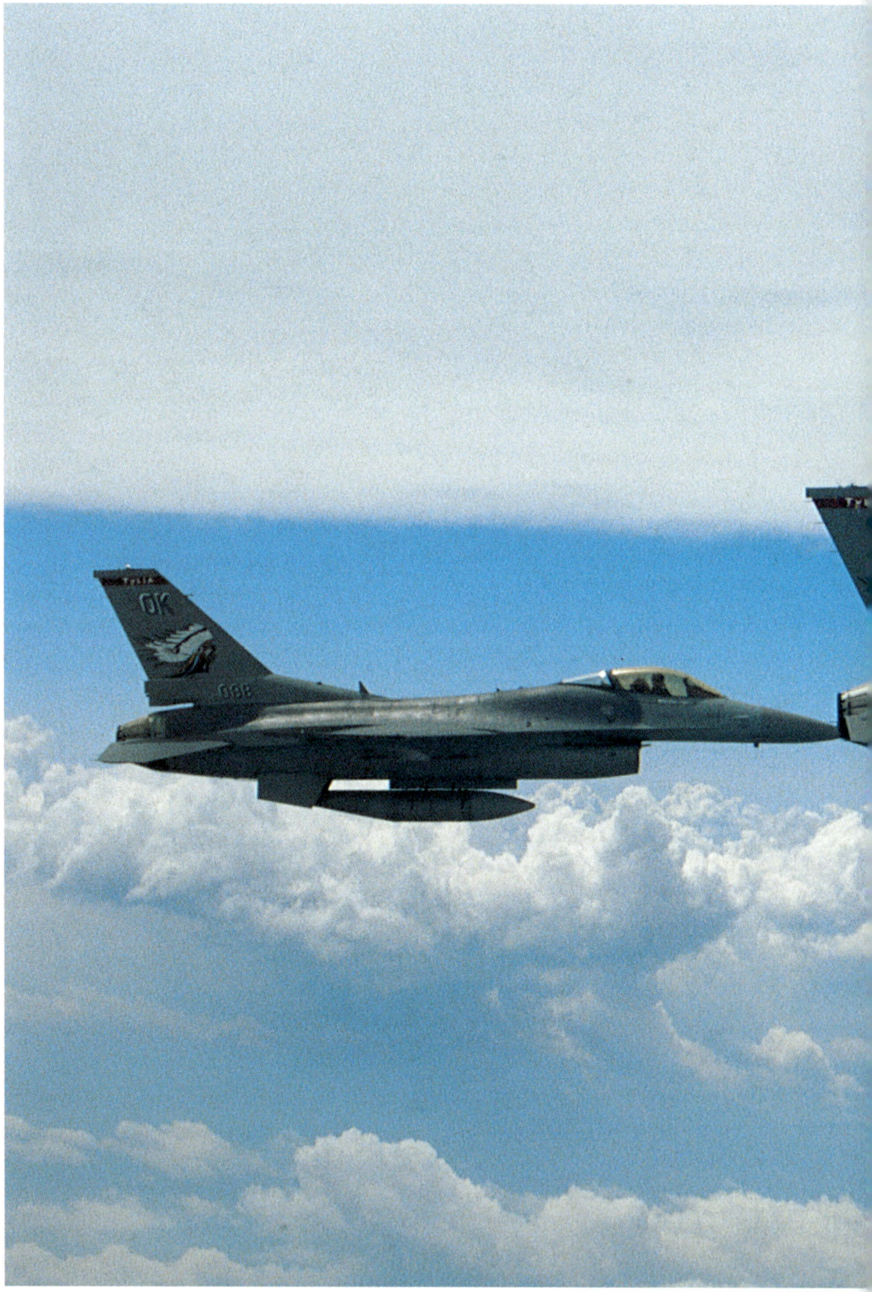

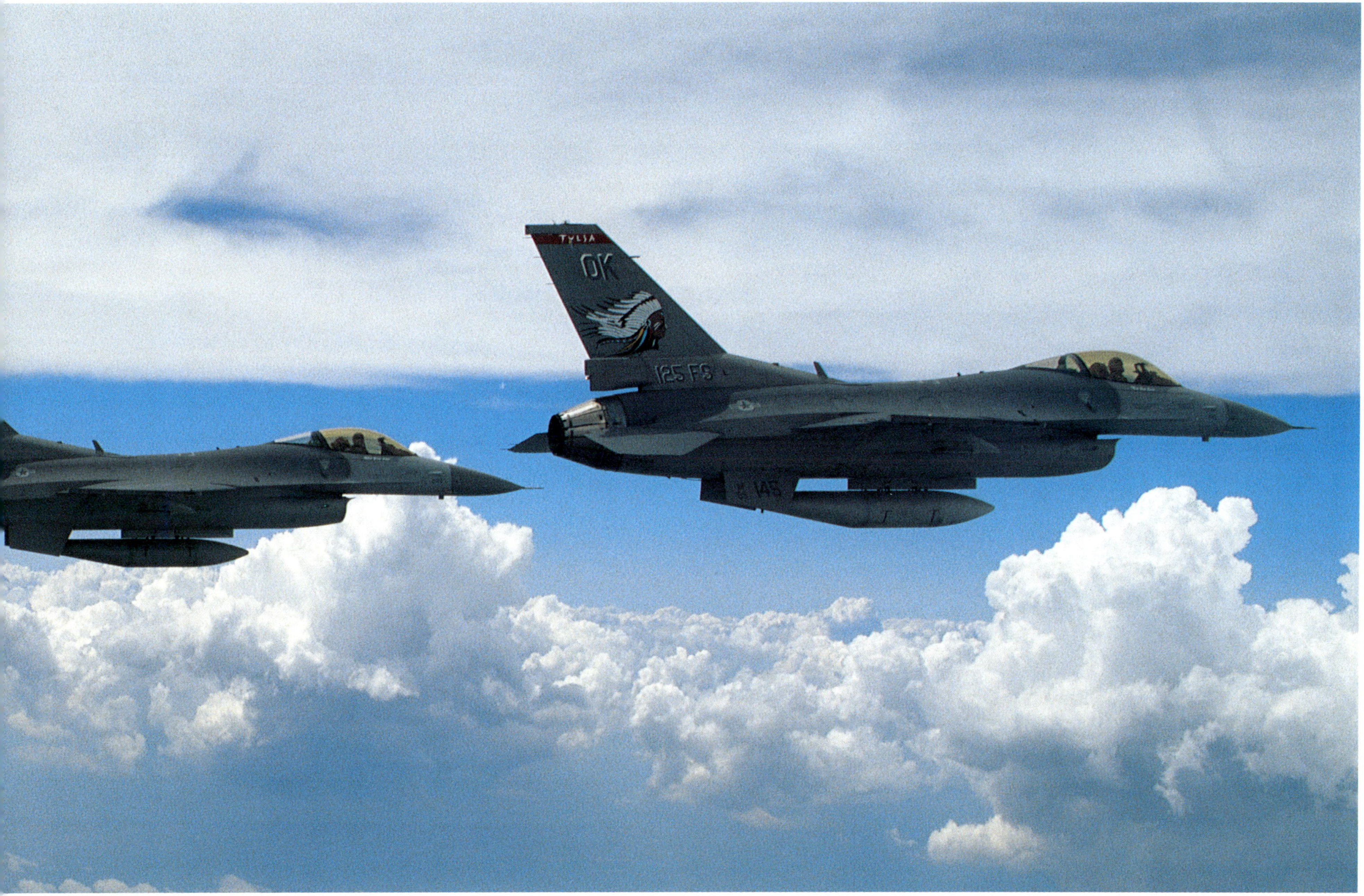
OK
125 FS
145

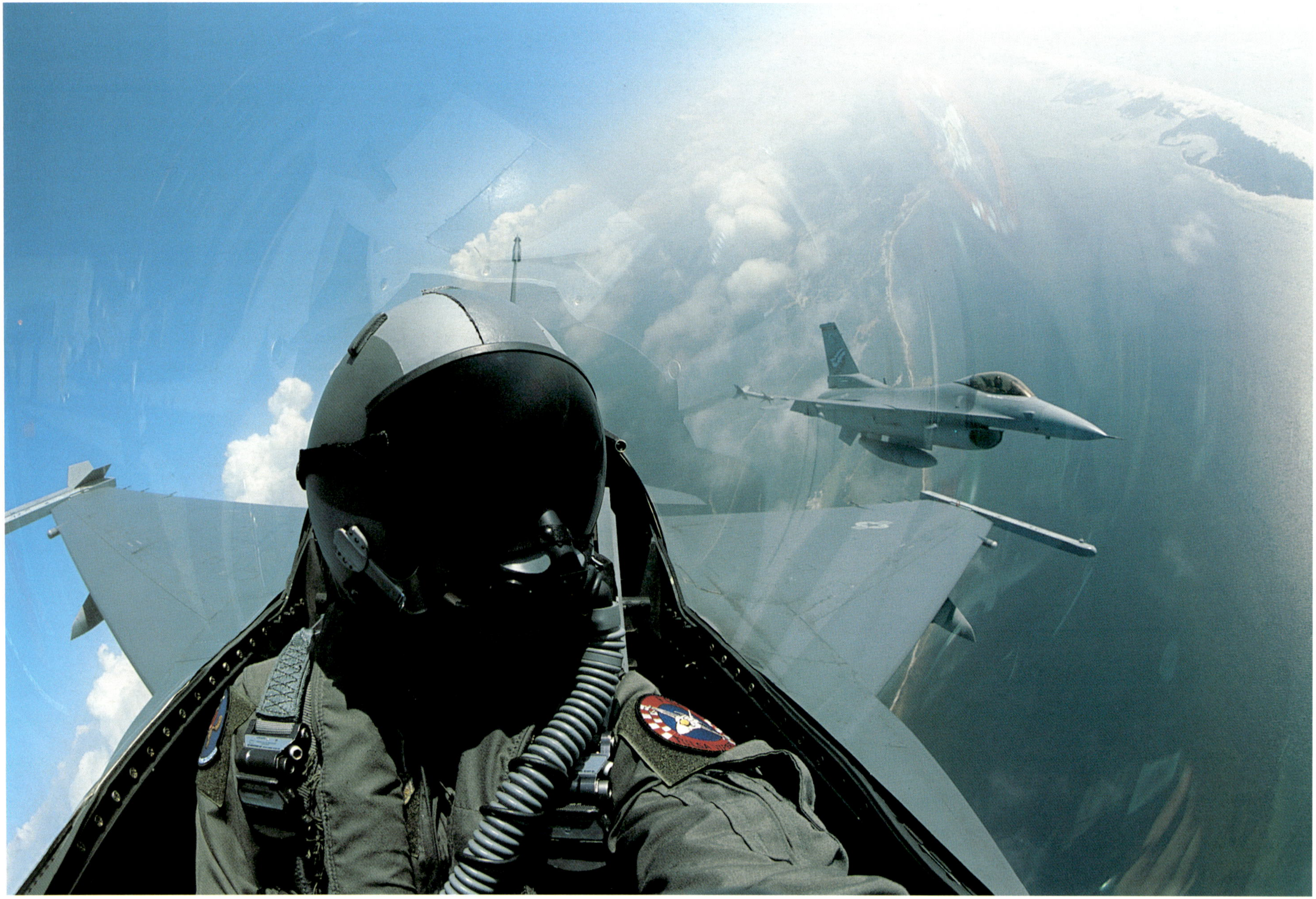

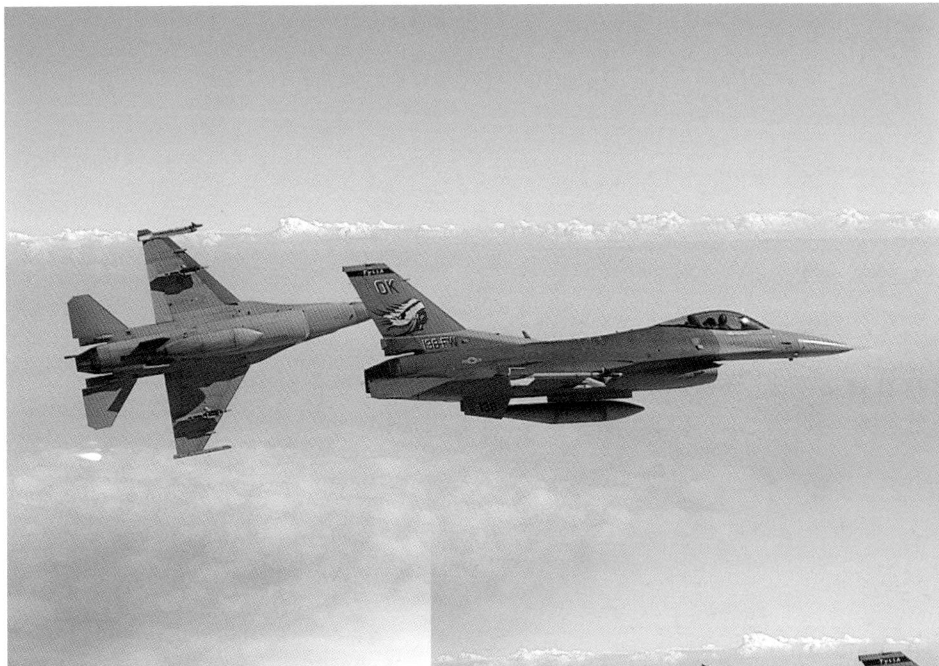

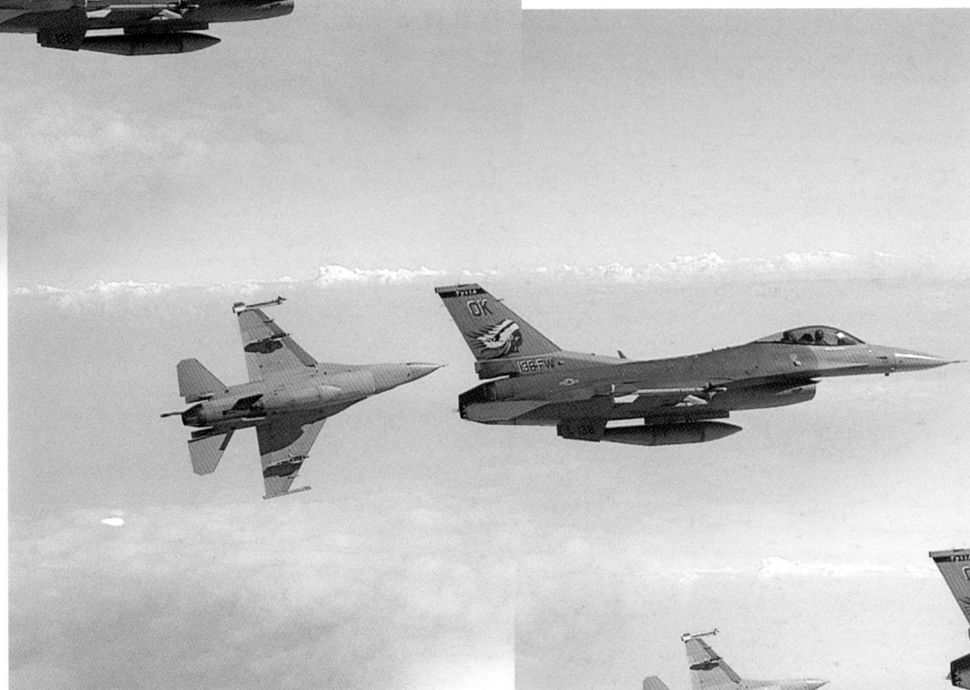

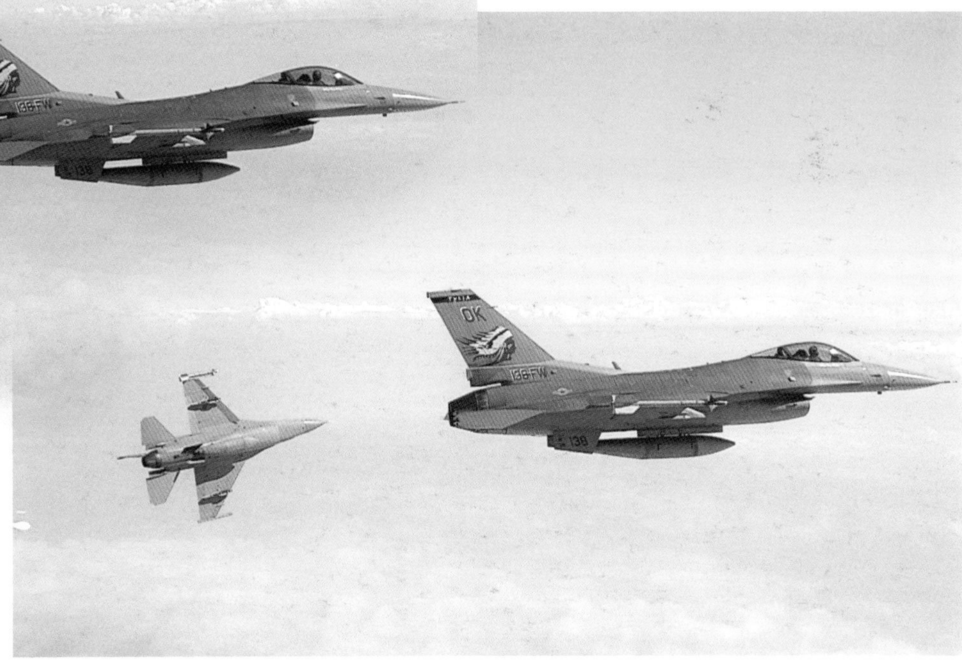

In this page
The two-ship formation is splitted, and in the same time flares are launched, with the purpose to decoy IR-guided missile threats.

In the page overleaf
Close-in formation is practised also during ferry flights.

In the page on the left
A close two-ship formation is flown just after take-off from Tyndall AFB.

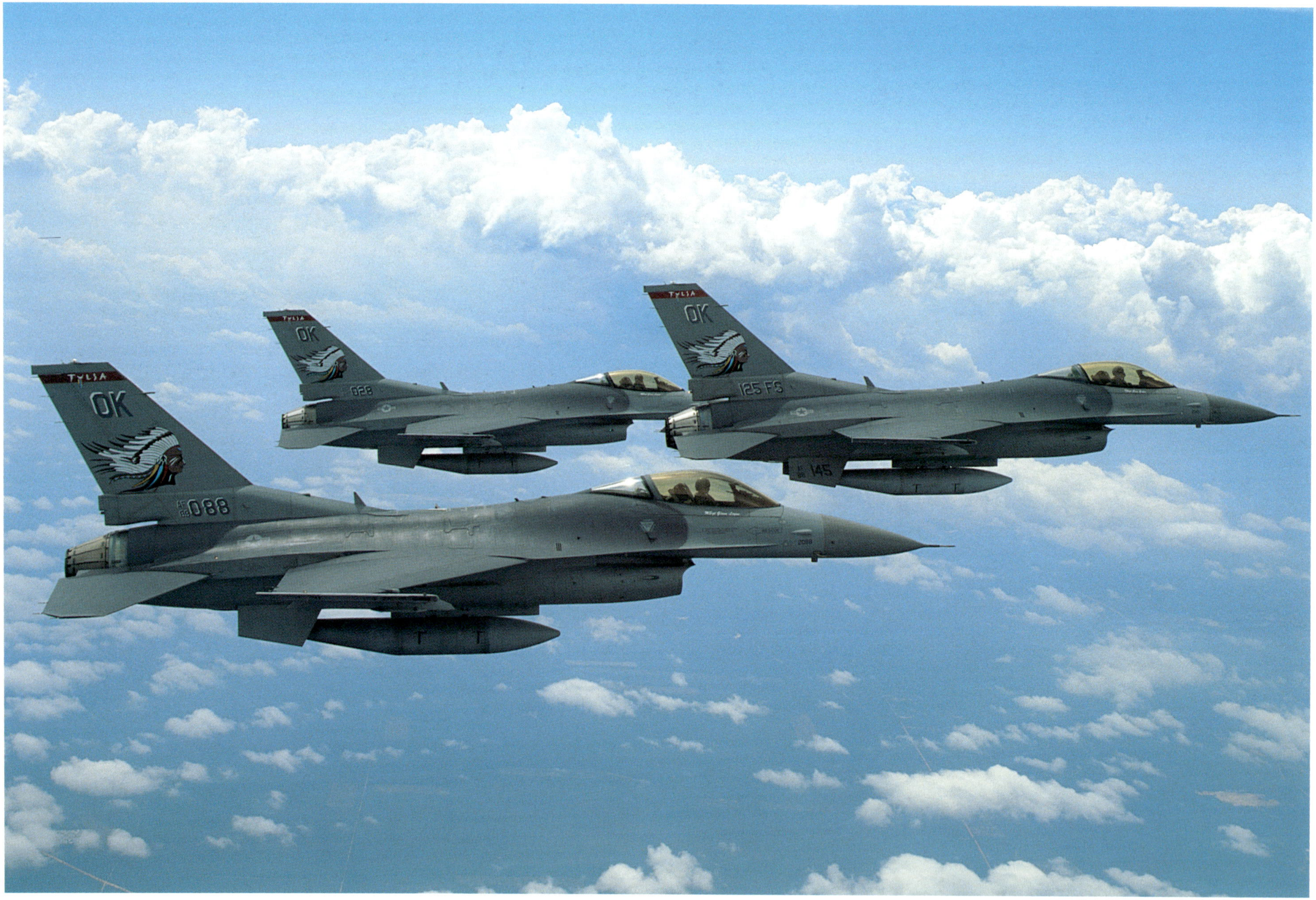
TULSA
OK
088
OK
028
TULSA
OK
125 FS
145

147th Fighter Wing

Texas

147th Fighter Wing, Texas (emblem approved 24 November 1982). The two shaded gold and yellow star is symbolic of the "Lone Star State" of Texas with golden opportunities available. The fortress design in the center of the emblem denotes the protection which the unit provides Texas and the defense of the United States. It also represents the bravery of unit members who died while defending their country. The ocean with waves and white caps are symbolic of the waters bordering the Texas coast which the unit protects.

111th Fighter Squadron, Texas (emblem approved 6 June 1933). The star represents Texas the "Lone Star State." Blue and orange are the U.S. Army Air Corps colors. The "Ace in the hole" is symbolic of the squadron's advantage over the enemy. The black and white border is symbolic of oil and cotton, the main mineral and agricultural assets of the state.

111th Fighter Squadron, Texas "Ace in The Hole". Organized at Kelly Field, Texas on 14 August 1917 as the 111th Aero Squadron, Air Service, this supply unit was redesignated as the 632d Aero Squadron on 1 February 1918. It was demobilized on 19 August 1919. In 1936 its lineage and honors were consolidated with those of the 111th Observation Squadron. On 29 June 1923, the 111th Observation Squadron was activated and extended federal recognition at Ellington Field, Houston, Texas. It was the aviation unit for the 36th Division, Texas National Guard, and was equipped with Curtiss JN-4B and JN-4H trainers. In December 1928 the unit moved to Houston Municipal Airport. From 1928 to 1940 it converted to different aircraft, and it was mobilized on 25 November 1940, in anticipation of America's entry into World War II. In the Fall of 1942 the squadron was shipped to North Africa. It flew anti-submarine warfare patrols, convoy duty, and border patrol missions with Douglas A-20As. The 111th Squadron was reorganized as a reconnaissance unit with A-36A and P-51 and F-6 versions of the North American Mustang.

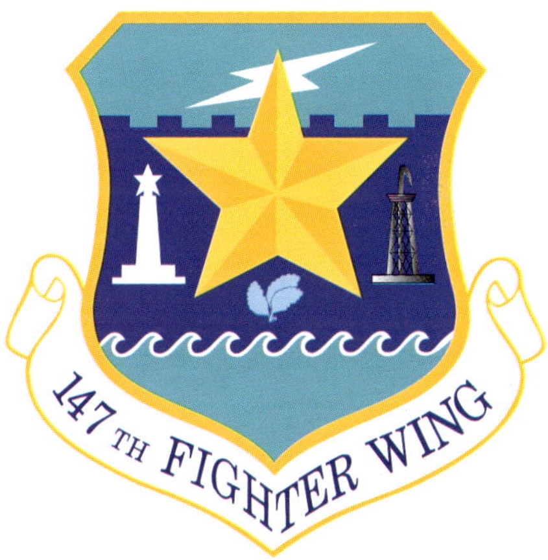

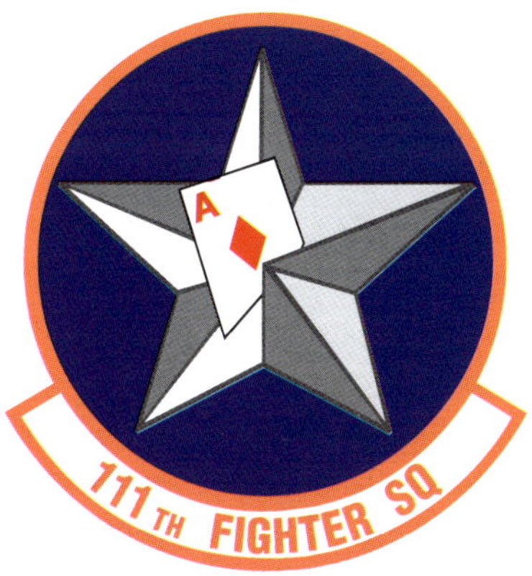

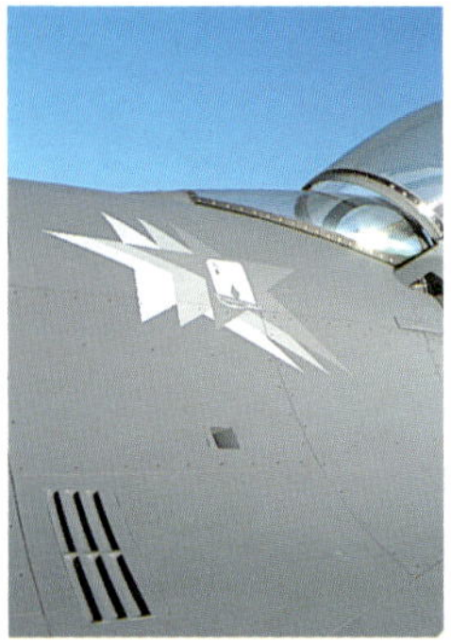
The Texas Lone Star.

The 111th Reconnaissance Squadron was assigned to the Twelfth Air Force and flew combat missions in the Mediterranean Theater of Operation (MTO) and European Theater of Operations (ETO) from July 1943 to Victory-Europe Day. It was inactivated in Creil, France in December 1945. The unit was redesignated as the 111th Fighter Squadron and allotted to the Texas National Guard on 24 May 1946. Extended federal recognition on 9 February 1947, the unit flew P-51D Mustangs at Ellington Field, Houston until it was mobilized for the Korean War on 10 October 1950. The unit was redesignated as the 111th Fighter-Bomber Squadron and converted to Republic F-84E Thunderjets. The unit moved to Itazuke AB, Japan in May 1951. From July 1951 to July 1952 the unit flew combat missions from Itazuke AB and then from Taegu AB, Korea. On 10 July 1952 the unit returned to state control at Houston municipal airport equipped with F-51Hs. The unit converted to North American F-86D Sabre Jets in August 1957. On 17 May 1958 the 147th Fighter Interceptor Group received federal recognition as the headquarters for the fighter squadron and supporting units.

With the implementation of the "gaining command" concept, the unit was assigned to the Air Defense Command (ADC) on 1 July 1960. The fighter squadron converted to Convair F-102 Delta Daggers in August 1960 and then added McDonnell F-101 Voodoos in May 1971 to become an ANG fighter training squadron and "Jet Instrument School".

On 1 April 1980, the unit was gained by Tactical Air Command following inactivation of ADC. The squadron converted to McDonnel F-4C Phantom IIs in fiscal year 1981. The unit converted to the General Dynamics F-16 Fighting Falcon in fiscal year 1991. The squadron designation changed to 111th Fighter Squadron on 15 March 1992 and became Air Combat Command gained on 1st June 1992. On 1 October 1995, the 147th Fighter Group was redesignated as the 147th Fighter Wing.

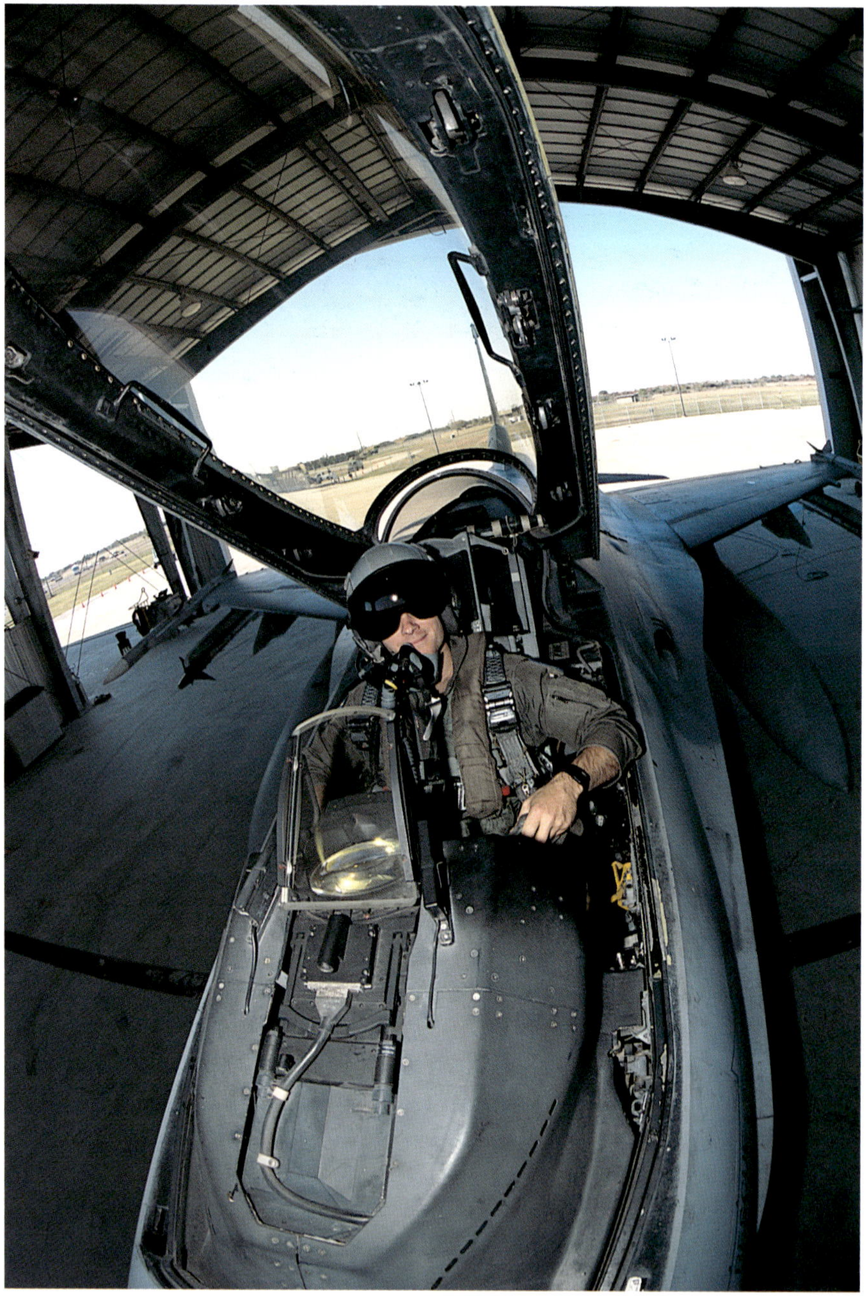

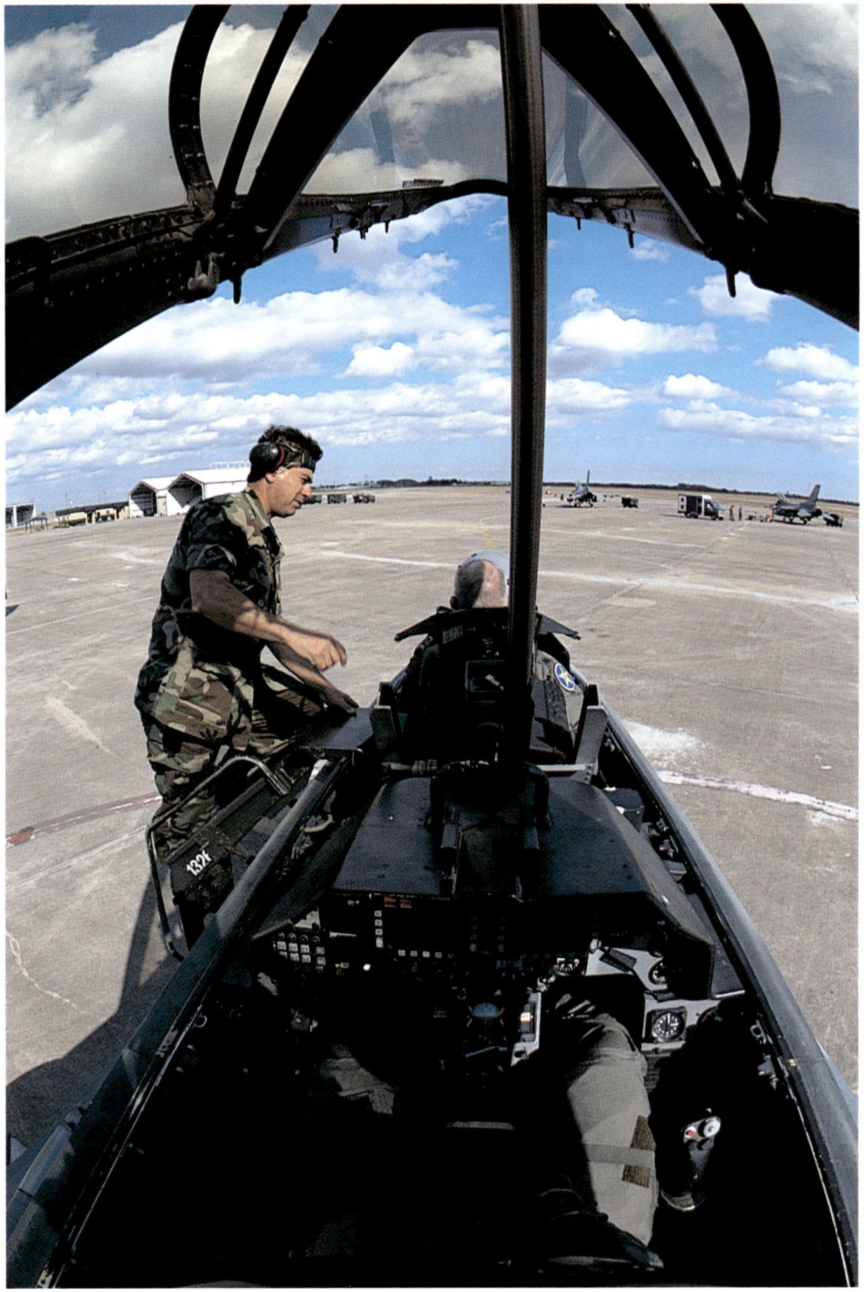

In the page on the left
Capt. John Birman in one of the two F-16Cs Block 25, ready for a scramble take-off.

On the left
Lt. Col. Steve "Snuffy" Higgs eases himself in the front seat of an F-16D, assisted by his crew chief.

On the right
A weapons specialist checks an AIM-9L Sidewinder IR-guided missile before a training sortie.

Below
Two technicians change the left main landing gear wheel of a Viper.

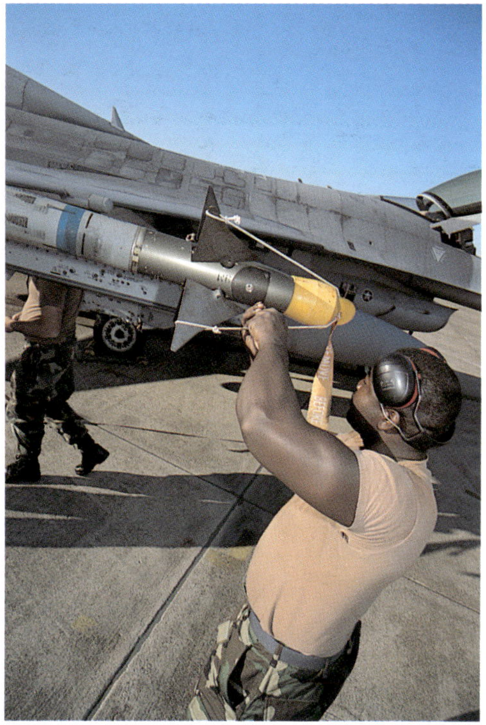

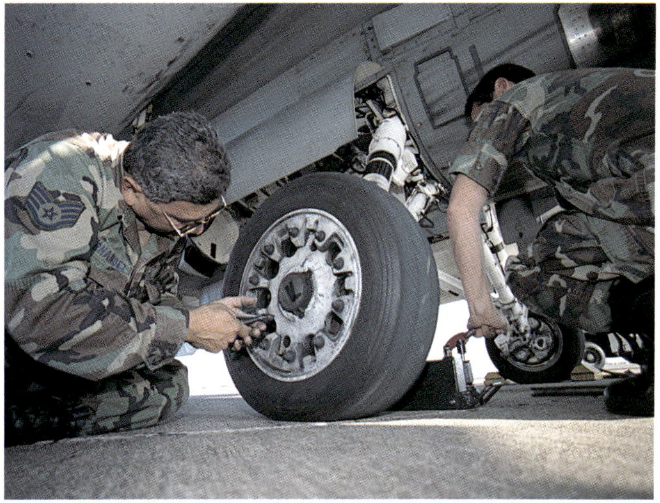

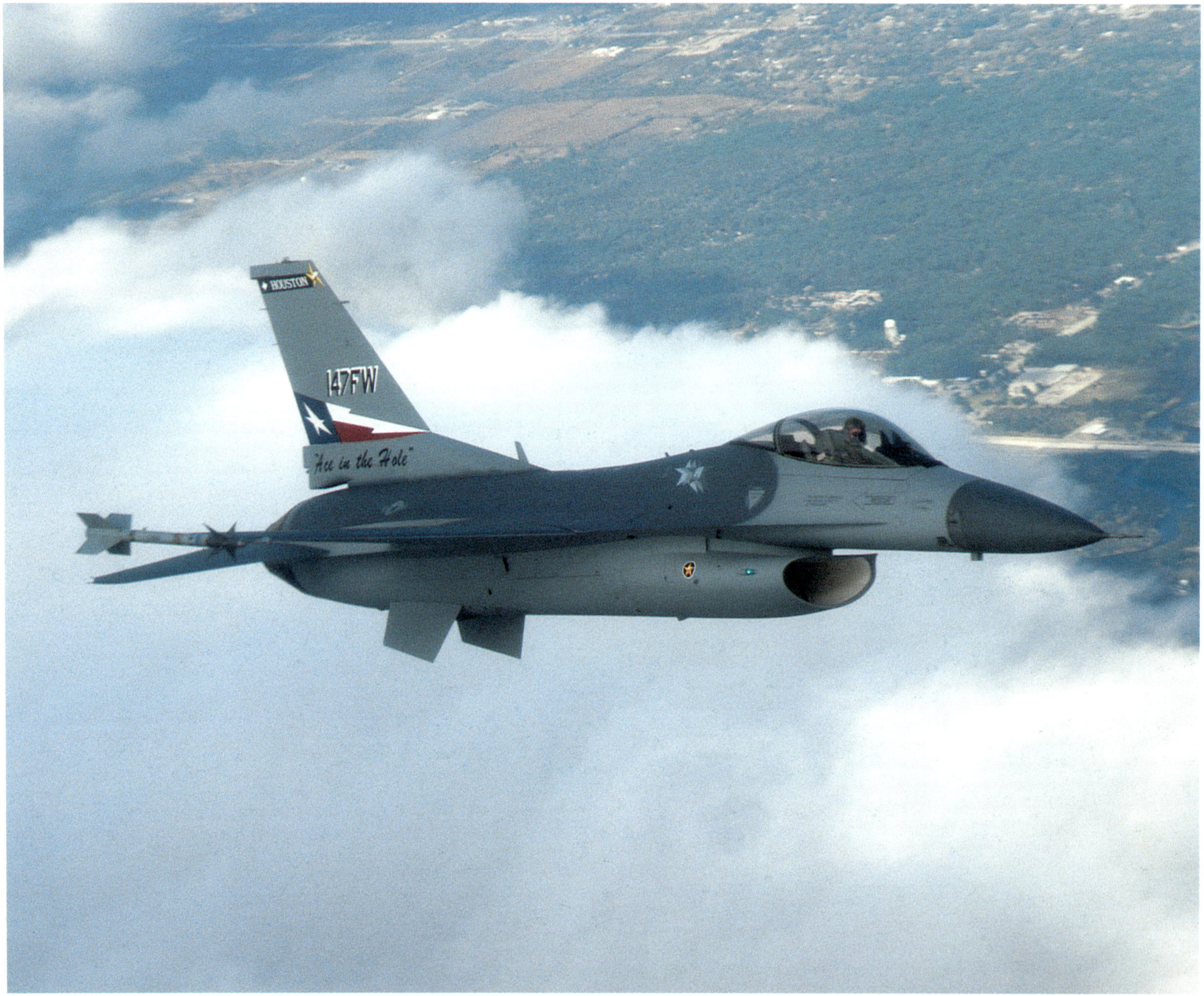

147th FW
FLIGHT LINE TALES
BY GIAMPAOLO AGOSTINELLI

"I've been flying the F-16 for twelve years: in practice since I started my career as a military pilot" says Capt. John Birman, one of the several 147th Fighter Wing highly experienced pilots with many hours in their flight log.
"I've logged over 1,500 flying hours in the Fighting Falcon A and C variants: an aircraft that did impress me a lot

On the left
The neat shape of the Fighting Falcon is highlighted in this close-up shot.

In the page on the right
A vertical climb over the Gulf of Mexico.

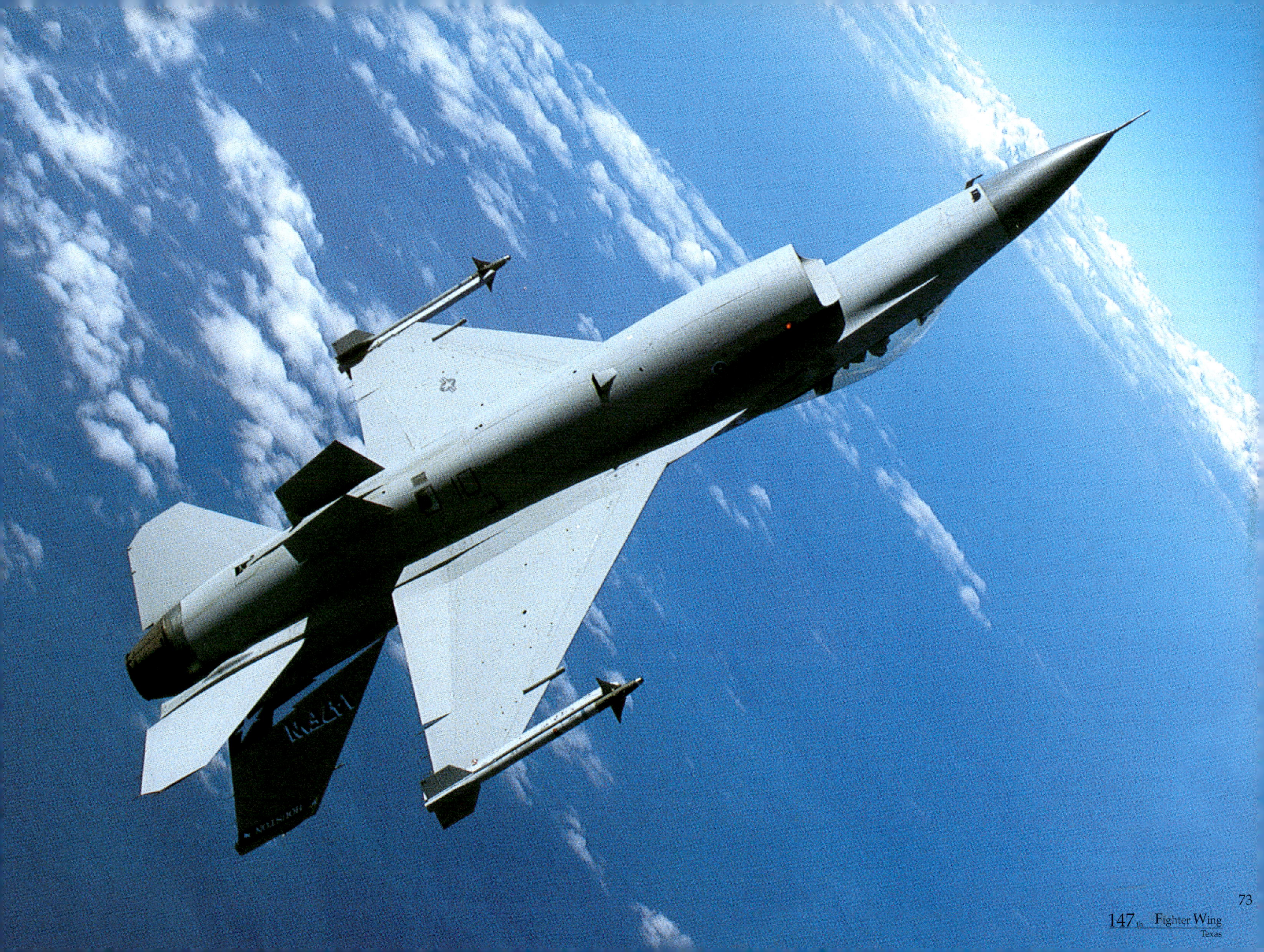

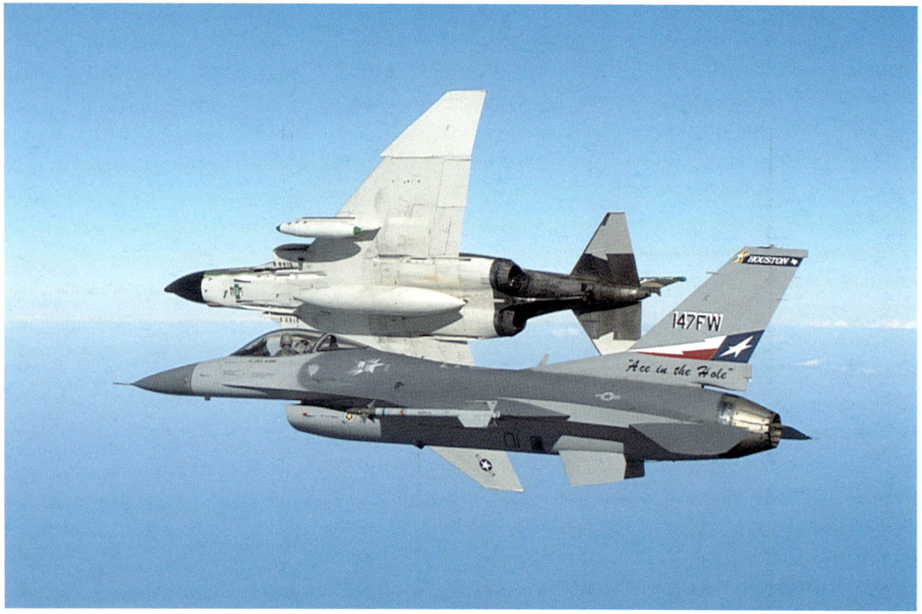

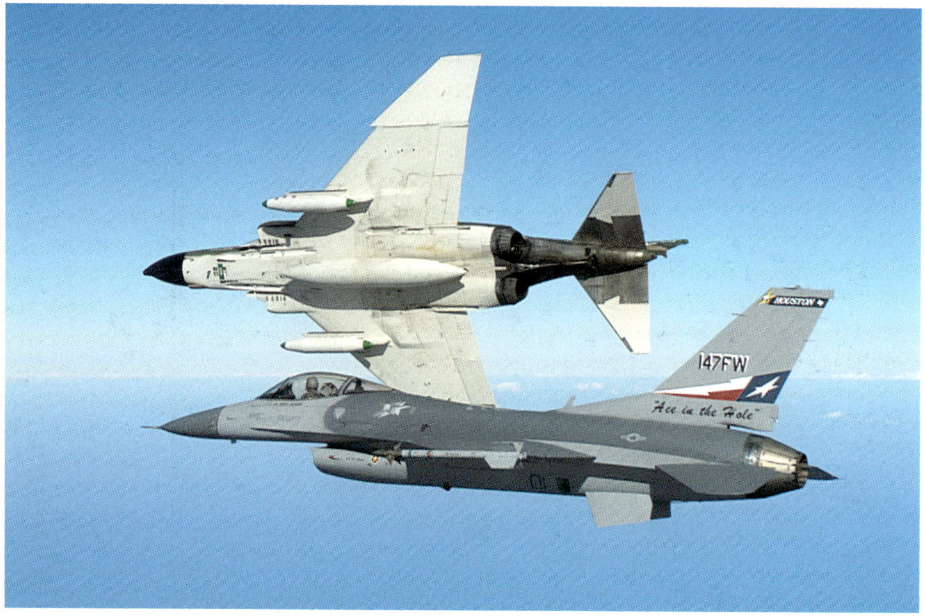

since the first time I flew the F-16, and of which one never ends discovering new characteristics and capabilities.

"As many turns as you wish and generous manoeuvrability are Viper's highlights. The fact is that you don't have to do or to move anything: you've got only to think in which direction you want it to go. And it goes!

"It was in 1996, we were deployed at Key West NAS in the deep South of Florida in order to ensure a ready-reaction force by means of continuous CAPs, Combat Air Patrol missions, and to guarantee armed protection in the case it was needed by maritime search and rescue operations in front of Cuba. One day that we were on a CAP mission with a pair of F-16Cs, we were advised that an uncommon gathering of military traffic in the Northern Cuban airspace, just a few miles South of our position" John recounts.

"Quite unusually, as many as twenty of MiG 29s and MiG-23s were aloft at the same time, and we knew we were at easy reach for them.

"The situation was surely very particular and such to raise our curiosity, but also suspect, and certainly to be handled with caution. We did not approach, we did not engage combat, but we simply acted in a way that they would be aware that we were controlling them on our radars. Consequently, they were able to control us as well. We were ready for action, to take over, is our commands would order us to do so, but all turned out like 'showing our muscles' from distance, to control and spy each other.

On top
A mid-air encounter with a vintage 555th FS Phantom II - past and present of air defence aircraft.

In the page on the right
The Phantom completes the break. The maneuver looks like the takeover in a relay race.

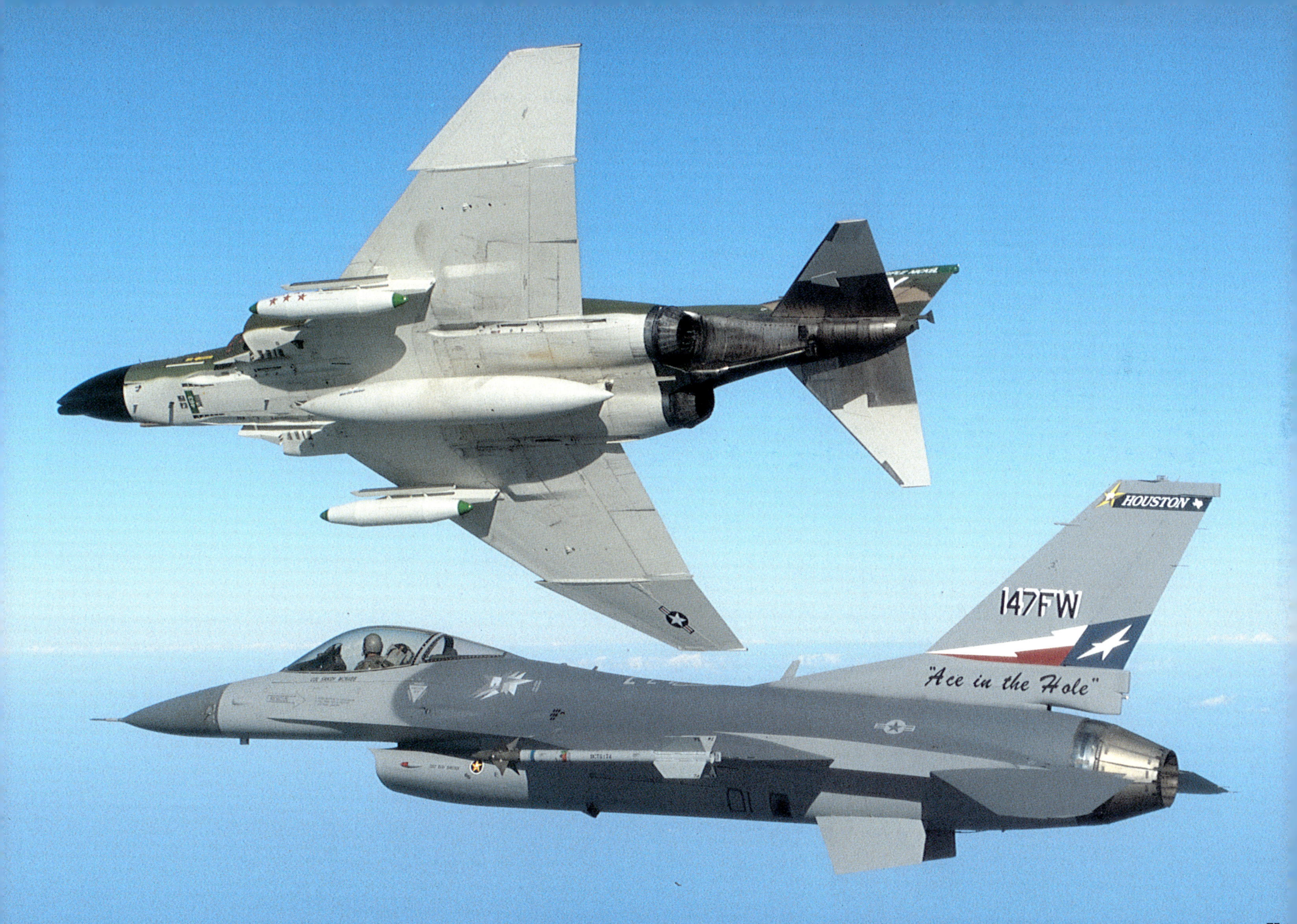
HOUSTON
147FW
"Ace in the Hole"

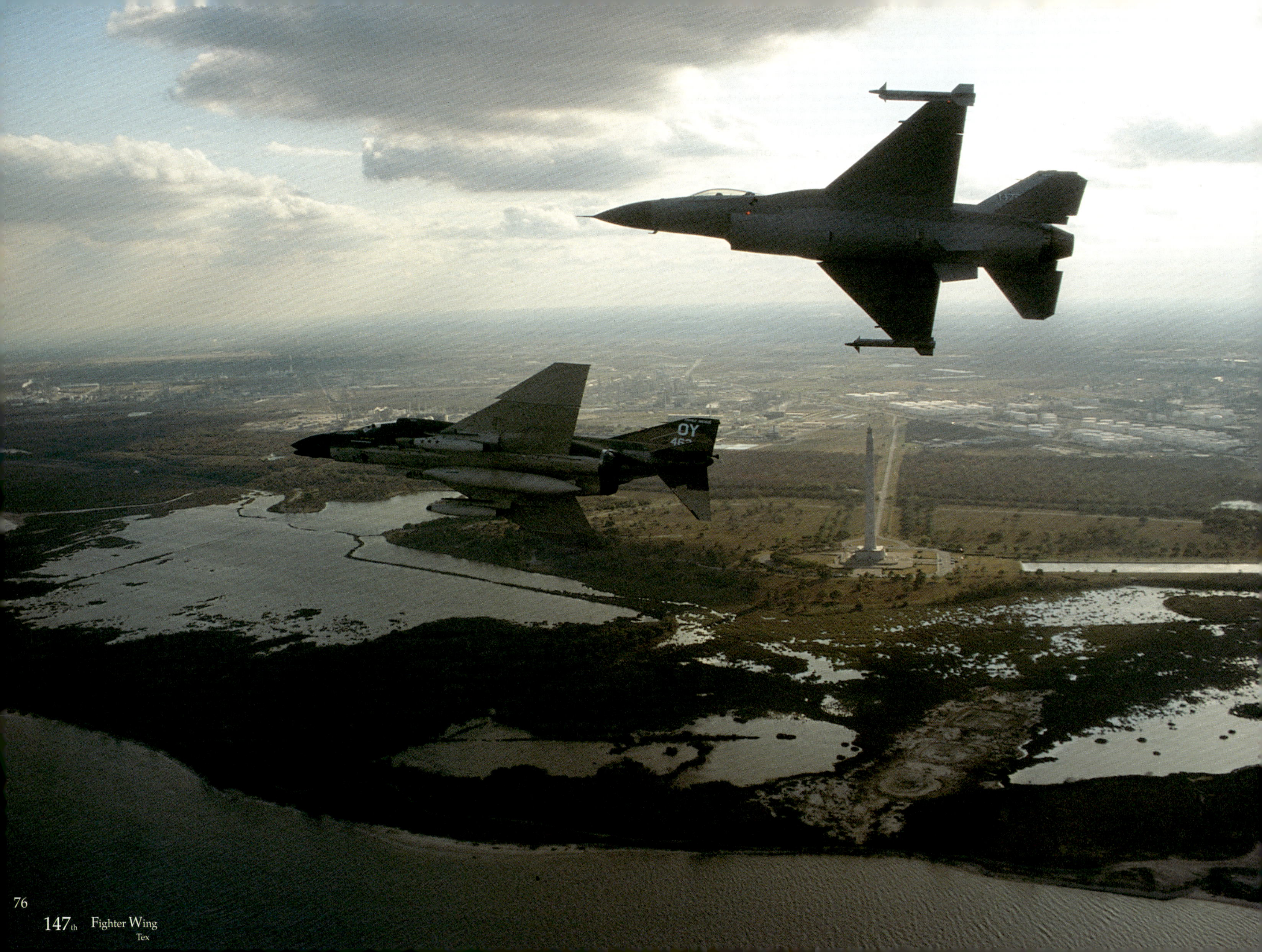
OY

In the page on the left
Turning over the Houston Obelisk, a memorial of Texas independence.

"Being the 147th Fighter Wing an air defense unit and also a branch of the 1st Air Force tasked with the protection of the continental United States airspace, we find ourselves in delicate situations fairly often. But whatever the situation we are to face, flying in an F-16 is a guarantee of sucess for us!" Birman concludes.

Sargeant Peter Hoffman, a weapons specialist, is a young man full of enthusiasm for his job, and in a special way for the Fighting Falcon. He has been working at the Weapons Shop for over three years and he is responsible for the training of new personnel.

"We bear a great responsibility because the 'youngsters' are very much influenced by the way we manage their training, which must be anyway brought to the highest standard" Peter explains.

"We must be fast, efficient and precise in weapons handling, as the correct and reliable operation of our weapons is a component of paramount importance, in case of real need. But also reloading our Vipers in due time is not a secondary goal", he adds.

"A major part of our training is devoted to reloading the weapons on the aircraft that re-enter from real missions, because they are needed to be ready for the next sortie in the shortest, possible time.

During one of these training drills, something bad was about to happen: one of those big mishaps which are always threatening us, if we don't pay more than enough attention.

We were a crew of two, and both of us were busy with loading a Sidewinder air-to-air missile: I was behind my colleague, instructing and guiding him to insert the missile properly in its rail on F-16 the wingtip, where now an AIM-120 AMRAAM is fitted.

We had a misunderstanding, at a certain point, and the result was that the Sidewinder was inserted only in a minimum stretch of its rail, and it was dangerously and menacely out of its housing. Luckily, the procedure reuquires us to cross-check positively our job frequently, before setting for the next step.

Therefore we noticed that inconvenience in time, and we were able to amend it quickly without slowing down or impairing other colleagues' jobs. When one has to act quickly, incidents are just around the corner; but permanent training allows us to improve always our performance, effectiveness and safety" Peter concludes.

Below
A 555th FS F-4D Phantom II leads a 147th FW F-16D over peaceful Texas waters.

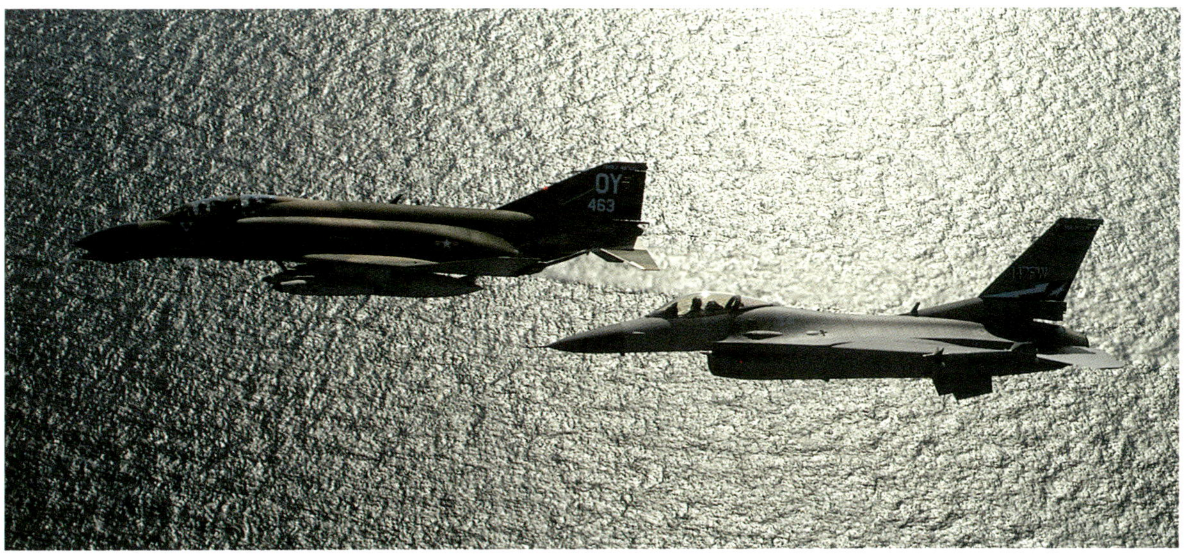

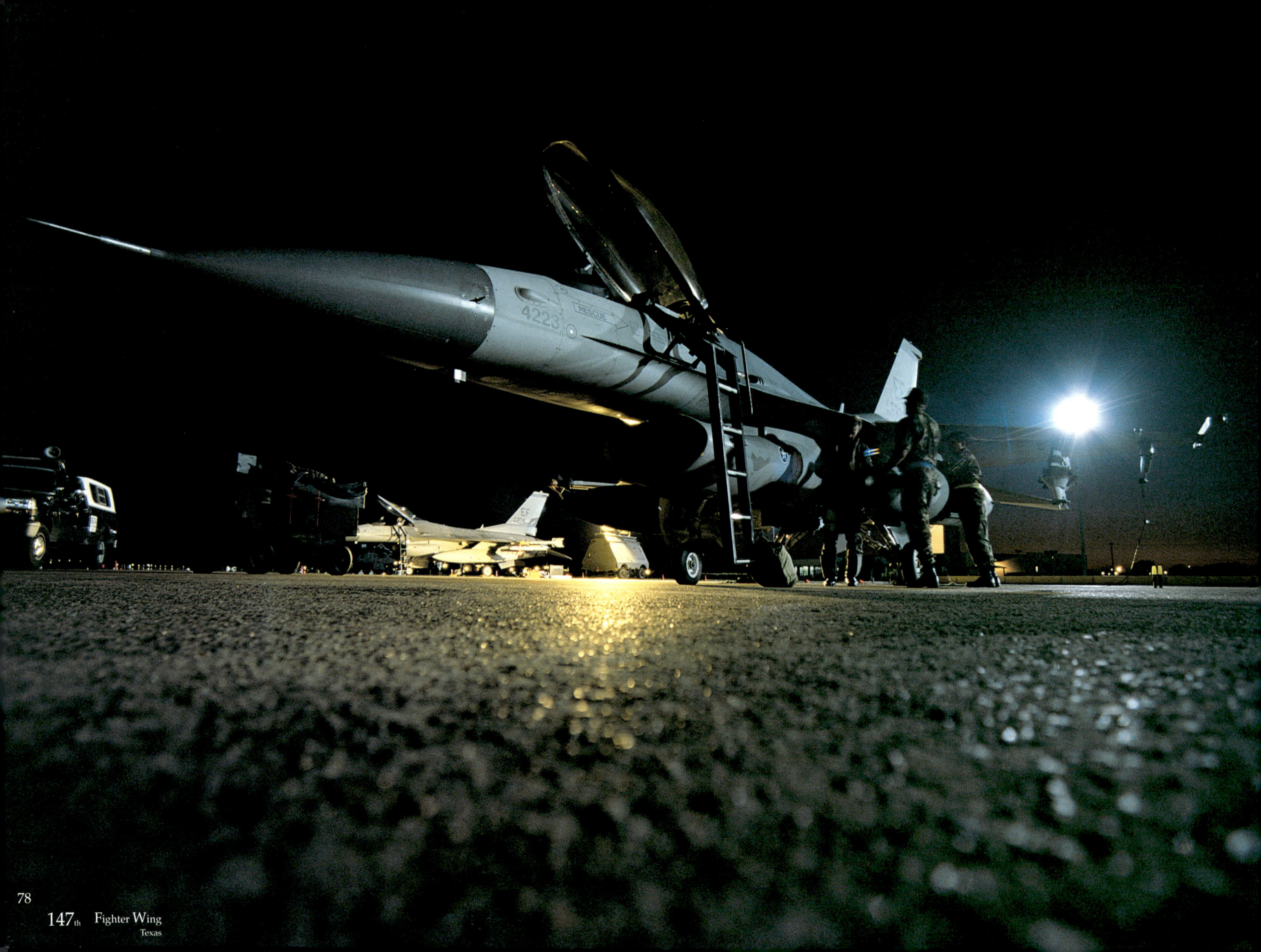
4223
RESCUE
EF

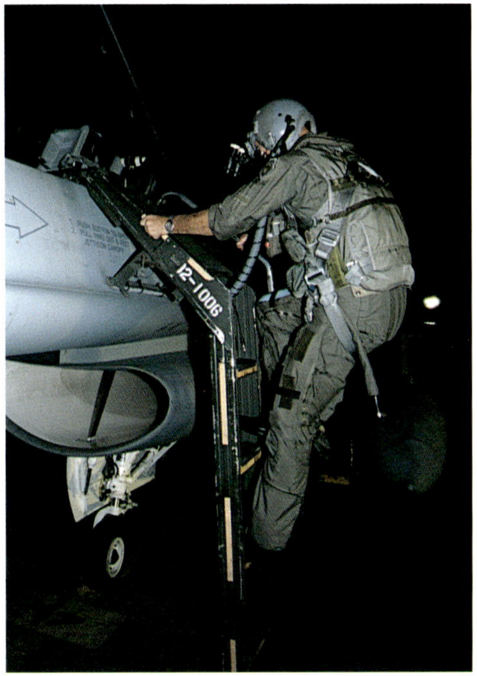

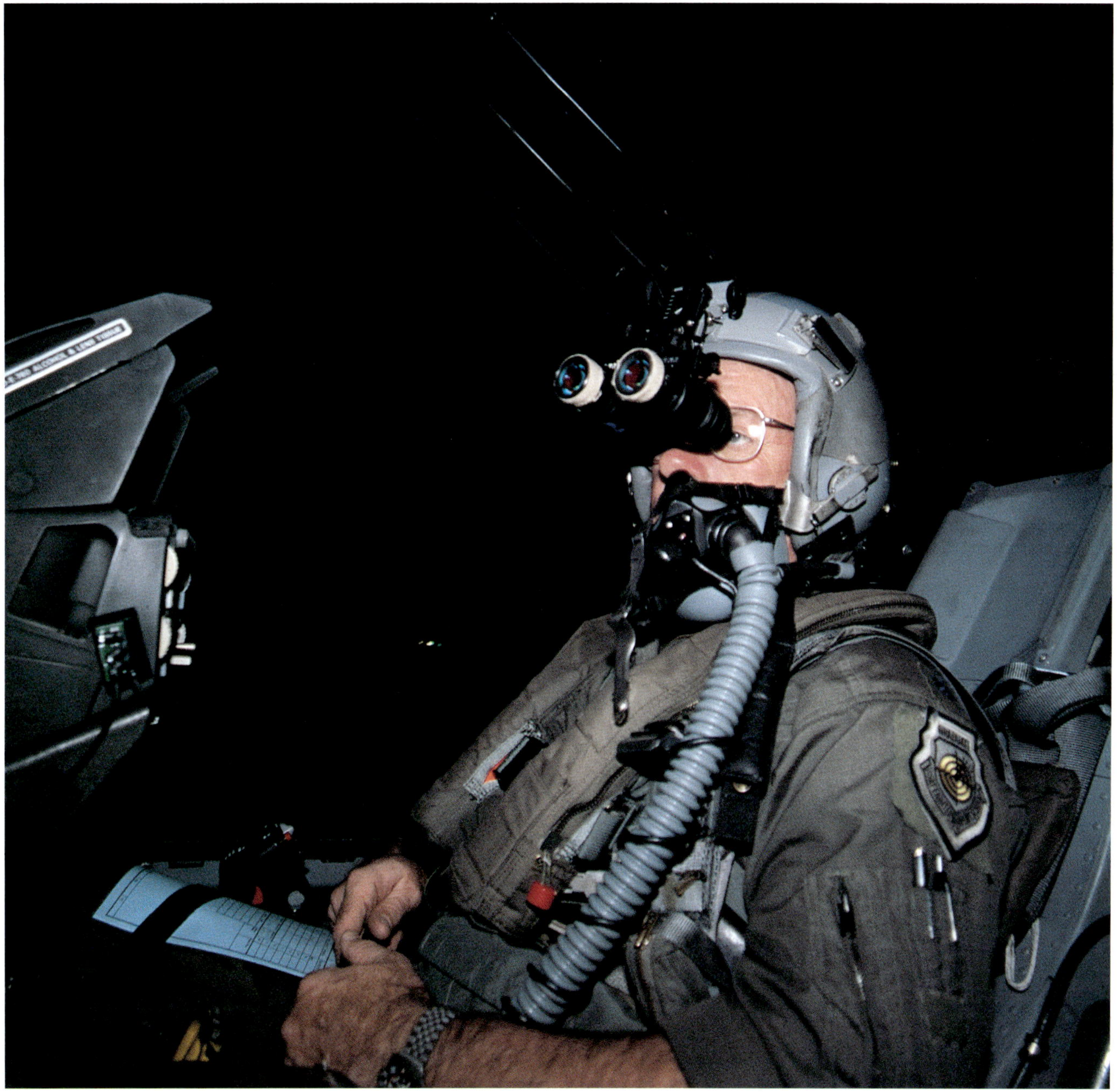

Above
Lt.Col. Steve "Snuffy" Higgs gets into his F-16D cockpit.

On the right
NVGs (Night Vision Goggles) are used in low-level tactical flight at night.

In the page on the left
Two 147th Figher Wing Vipers are prepared for a night training sortie.

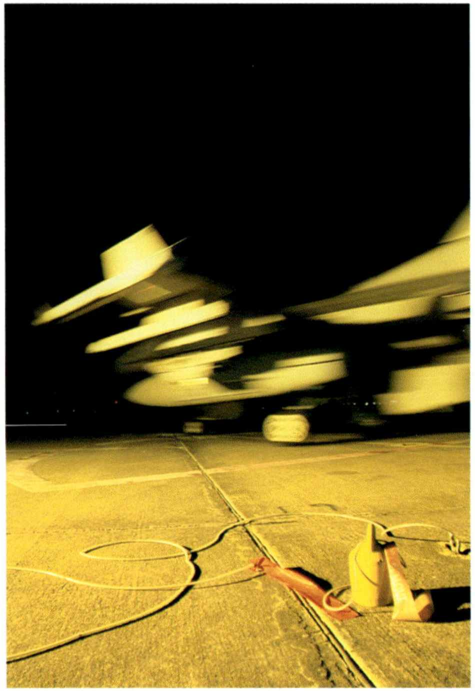

Above
This picture shows how the tarmac environment appears through the NVGs (3rd-generation ITT F-4949).

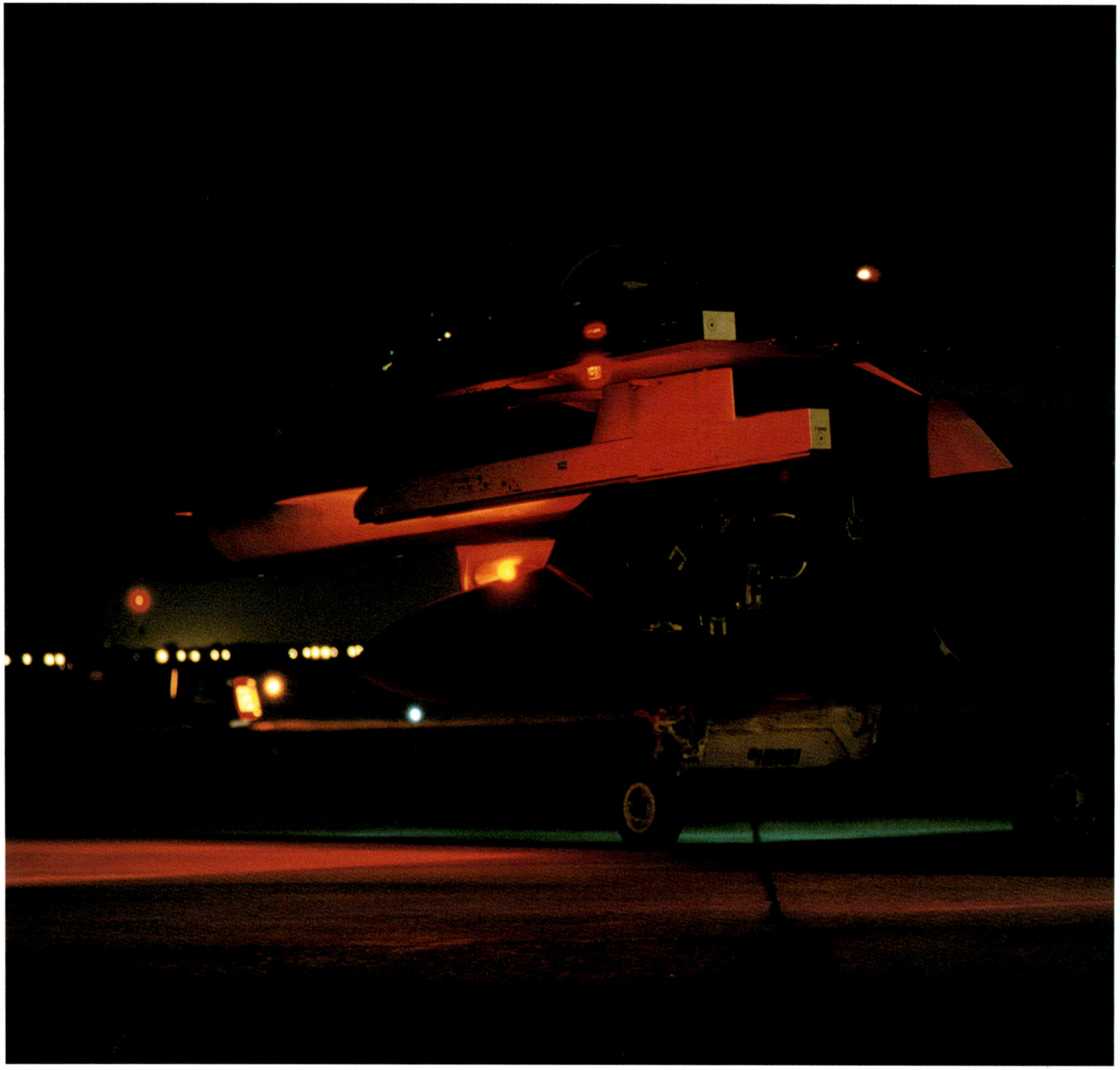

On the right
An F-16 taxiing at night, in a normal high-speed film shot.

148th Fighter Wing
Minnesota

148th Figher Wing, Minnesota (emblem approved 30 October 1964). The emblem signifies Polaris (the North Star) at its zenith in the Ursa Minor (Small Bear) constellation which refers to the State of Minnesota, the "North Star State." The winged arrowhead represents the airborne might of the unit's fighters. The Phoenix bird bears a bomb with nuclear symbol and lightning flashes in a cloud which denotes the overwhelming power and destructiveness of the weapons which the fighters can deliver. All elements represent the unit's mission to "Deter the Aggressor and Prevail in War."

179th Fighter Squadron, Minnesota (emblem approved 3 December 1953) The bulldog represents an aggressive and tenacious adversary in combat. The flying helmet and machine gun symbolize the basic mission of the unit which is Air Defense.

Being close to the border of Canada, the motto of the squadron is "Bulldogs of the Border."

179th Fighter Squadron, Minnesota "Bulldogs of the Border".

Constituted at Hamilton Field, California, in May 1943, the squadron was activated in July as the 393rd Fighter Squadron. It was assigned to the 367th Fighter Group, Ninth Air Force.

The squadron flew combat operations with Lockheed P-38Js in the European Theater of Operations beginning in May 1944. The squadron converted to Republic P-47D Thunderbolts in February 1945 and continued to fly combat missions until the end of the war. The squadron was inactivated at Seymour Johnson Field, North Carolina, in November 1945.

The unit was reconstituted and redesignated as the 179th Fighter Squadron and allotted to the Minnesota National Guard on 24 May 1946. On 18 September 1948, the squadron received federal recognition at Duluth Municipal Airport. In December 1948 the unit was equipped with North American P-51D Mustangs and support aircraft. The squadron was mobilized for

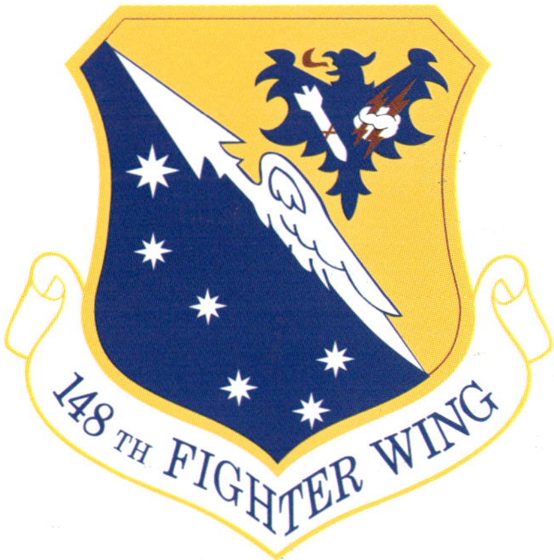

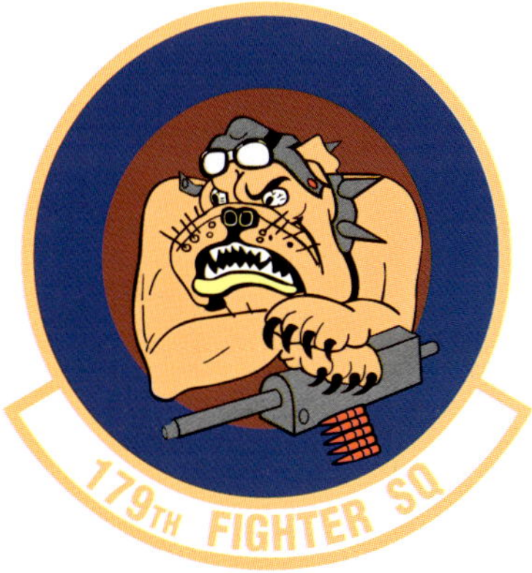

the Korean War on 1 March 1951, but remained at Duluth returning to state control on 1 December 1952. The unit converted to Lockheed F-94A/B Starfires and to F-94Cs in August 1957. In July 1959 the unit converted to Northrop F-89J Scorpions and became Air Defense Command gained on 1 July 1960 when the 148th Fighter Interceptor Group received federal recognition.

In November 1966 the unit converted to Convair F/TF-102A Delta Daggers and in April 1971 to McDonnell F-101B/F Voodoos. During Winter 1975 the unit was redesignated as the 179th Tactical Reconnaissance Squadron and converted to RF-4C Phantoms. On 15 November 1983 the unit was redesignated as the 179th Fighter Interceptor Squadron and converted to F-4D Phantoms.

From April 1986 to April 1987 the unit provided aircraft and personnel for Operation *Creek Klaxton*, the ANGs assumption of the air defense mission at Ramstein AB, Germany.

This was done while the 86th Tactical Fighter Wing, US Air Forces Europe, converted to F-16C/D Fighting Falcons. In Winter 1987, the squadron sent a detachment to Loring AFB, Maine, to take over the air defense alert mission while the 101st Fighter Interceptor Squadron, Massachusetts ANG, converted to F-15A/B Eagles. During Summer 1989 the squadron sent a detachment to Fresno ANGB and George AFB, California, to take over the air defense alert mission from the 194th Fighter Interceptor Squadron while that squadron converted to General Dynamics F-16A/B air defense fighters. During Winter 1990 the squadron completed its conversion to General Dynamics F-16 A/B Fighting Falcons. On 15 March 1992 the unit was redesignated as the 179th Fighter Squadron. On 1 June 1992 the squadron's gaining command changed to Air Combat Command.

On the right A 148th Fighter Wing pilot taxies his F-16 ADF for a scramble take-off.

In the page on the right A Minnesota ANG Viper in a Tyndall AFB shelter, ready for scramble.

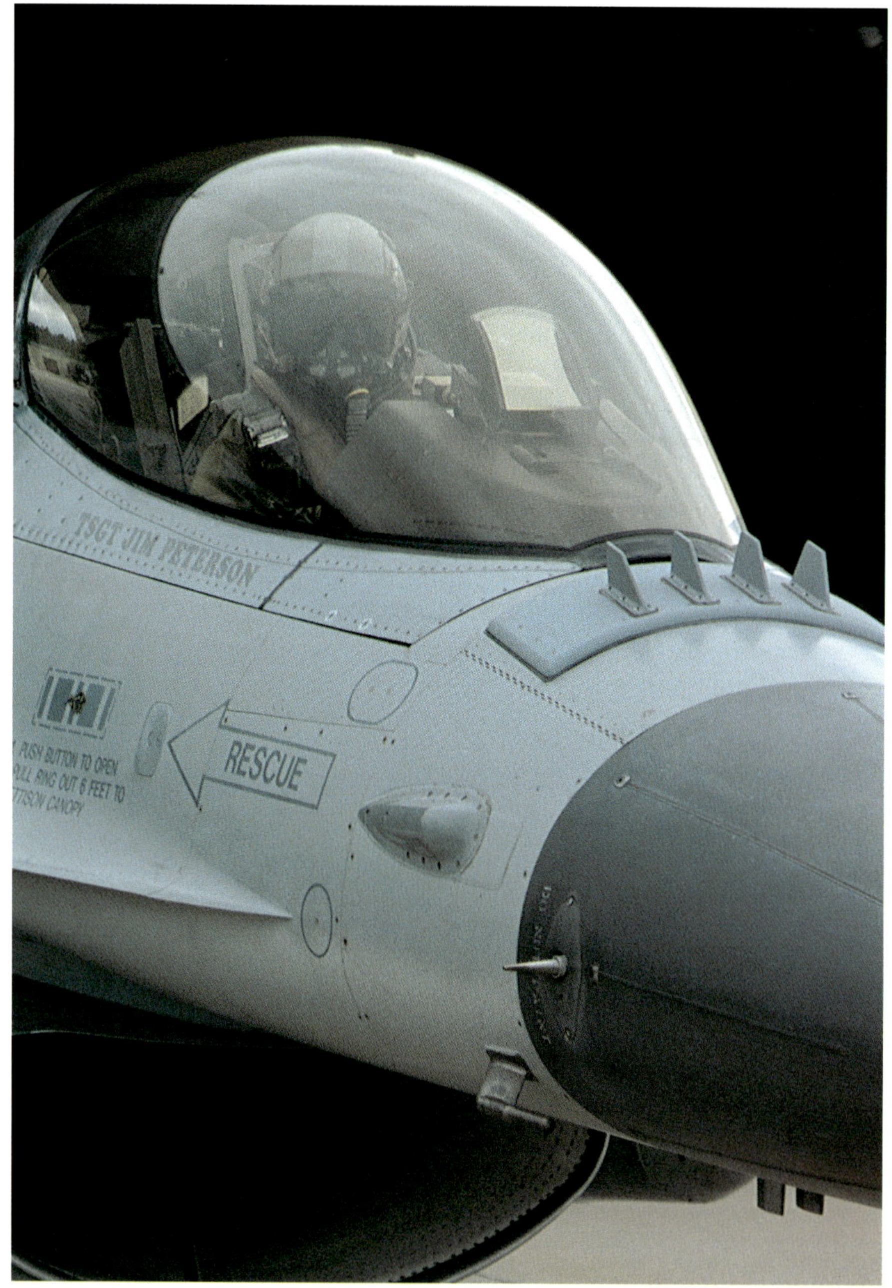

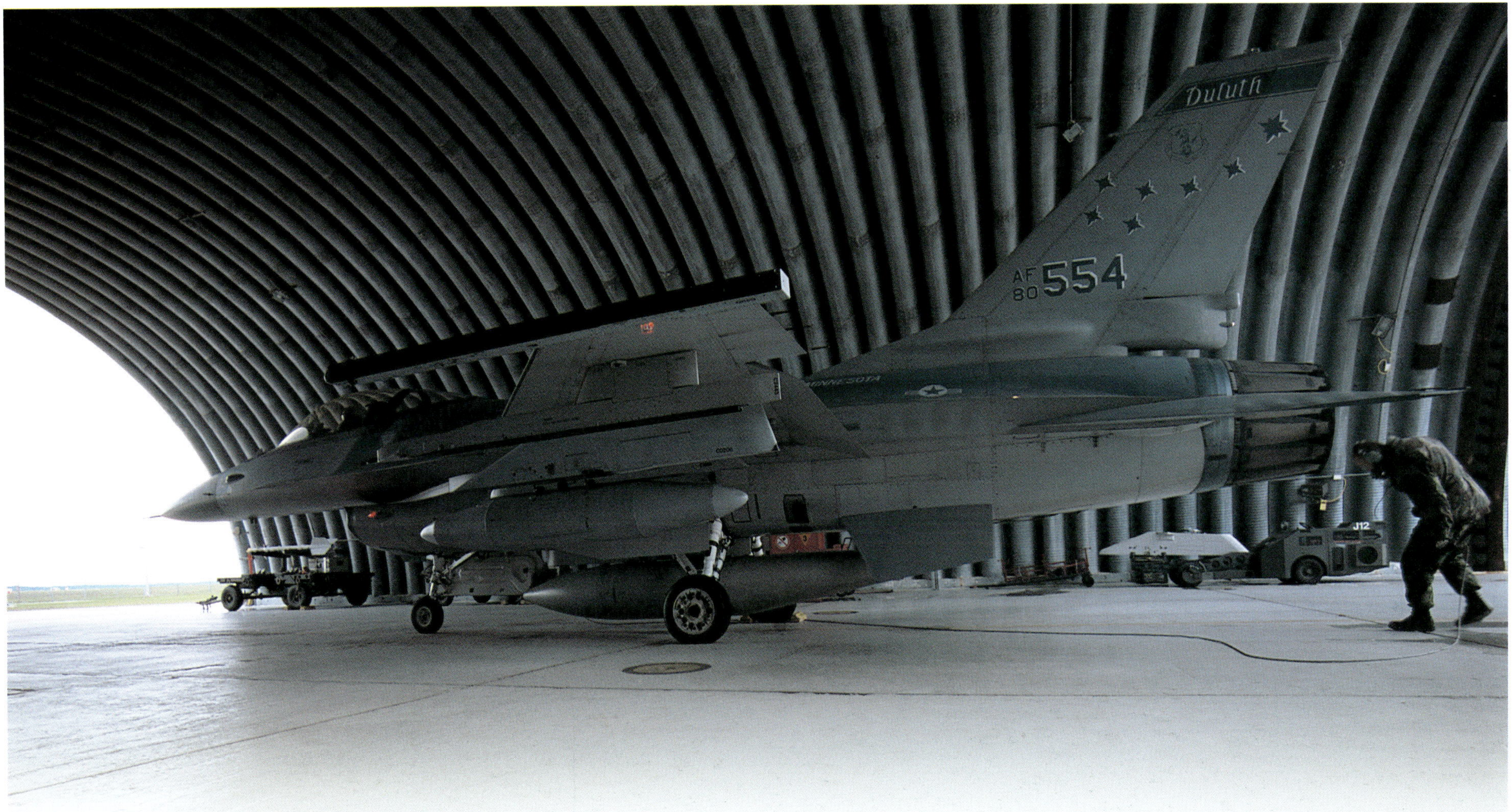

THE FIRST AIR FORCE AND THE AIR SOVEREIGNTY

BY MAJ. DONALD ARIAS

More than 200 years ago, the framers of the Constitution envisioned a "well-regulated militia" of citizen-soldiers standing ready for the defense of the homeland. Now, in the new millennium, the Guard has come full-circle again standing ready for our nation's defense at 1st Air Force. The responsibility for American air defense was transferred from the regular Air Force to the Air National Guard in October 1997. At that point, 1st Air Force became the first numbered air force to be made up primarily of citizen-airmen.

Major General. Philip G. Killey, previously the Director of the Air National Guard and the Commander during the transition, was the first to command 1st Air Force under this totally new and innovative concept.

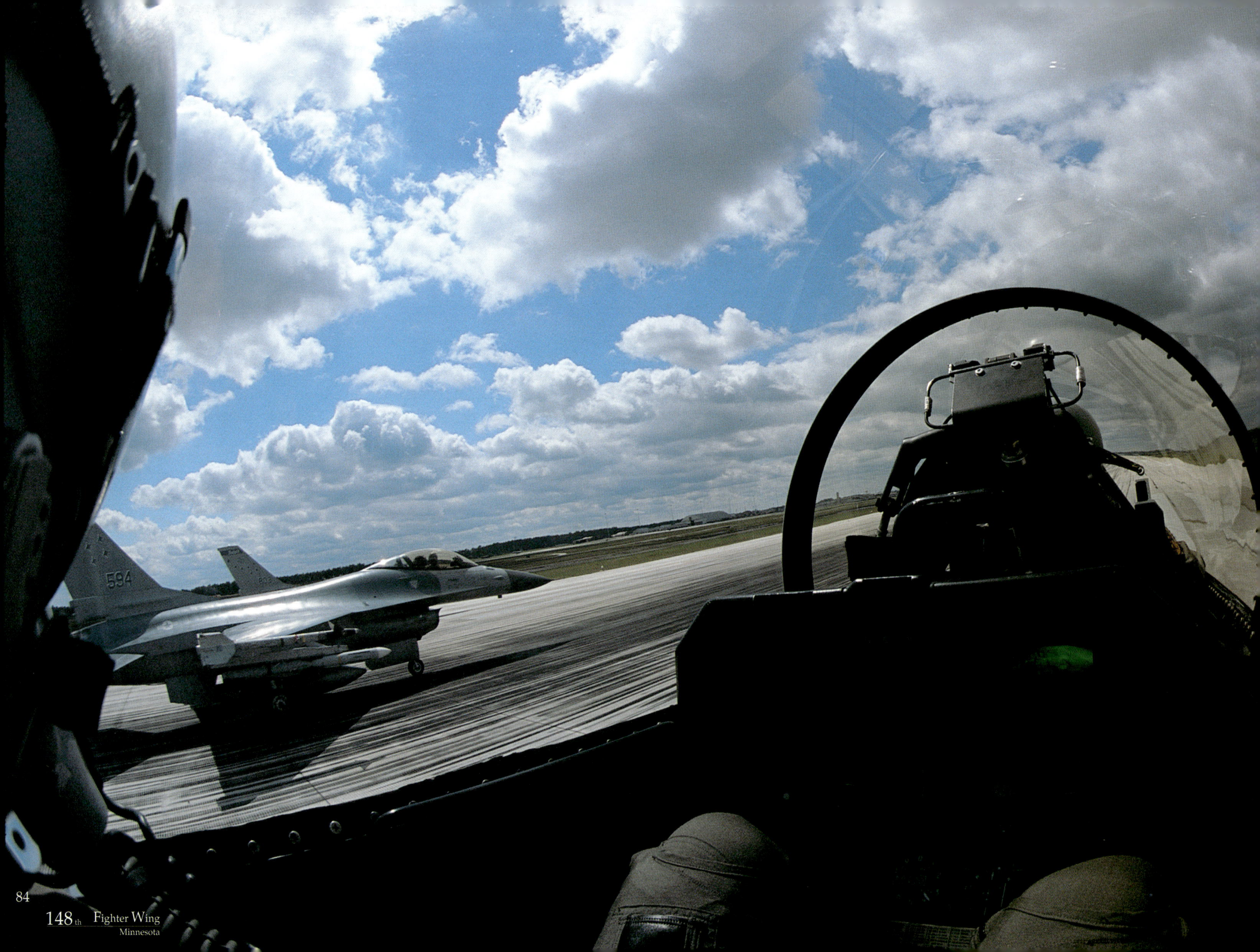
594

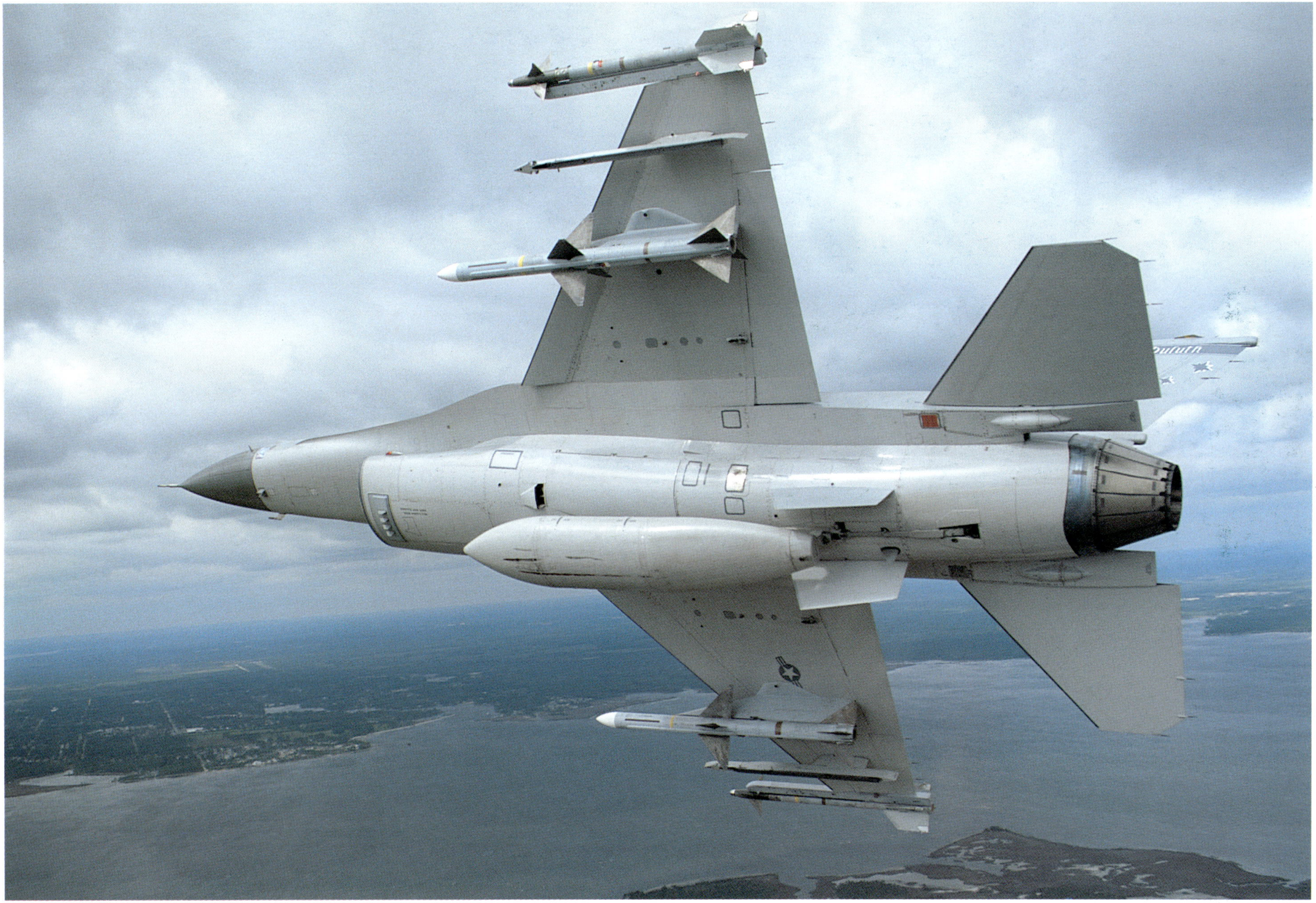

In the page on the left Brakes released for formation take-off.

Above Radar-guided AIM-7 Sparrow and IR-guided AIM-9 Sidewinder missiles arm the early F 16 ADFs.

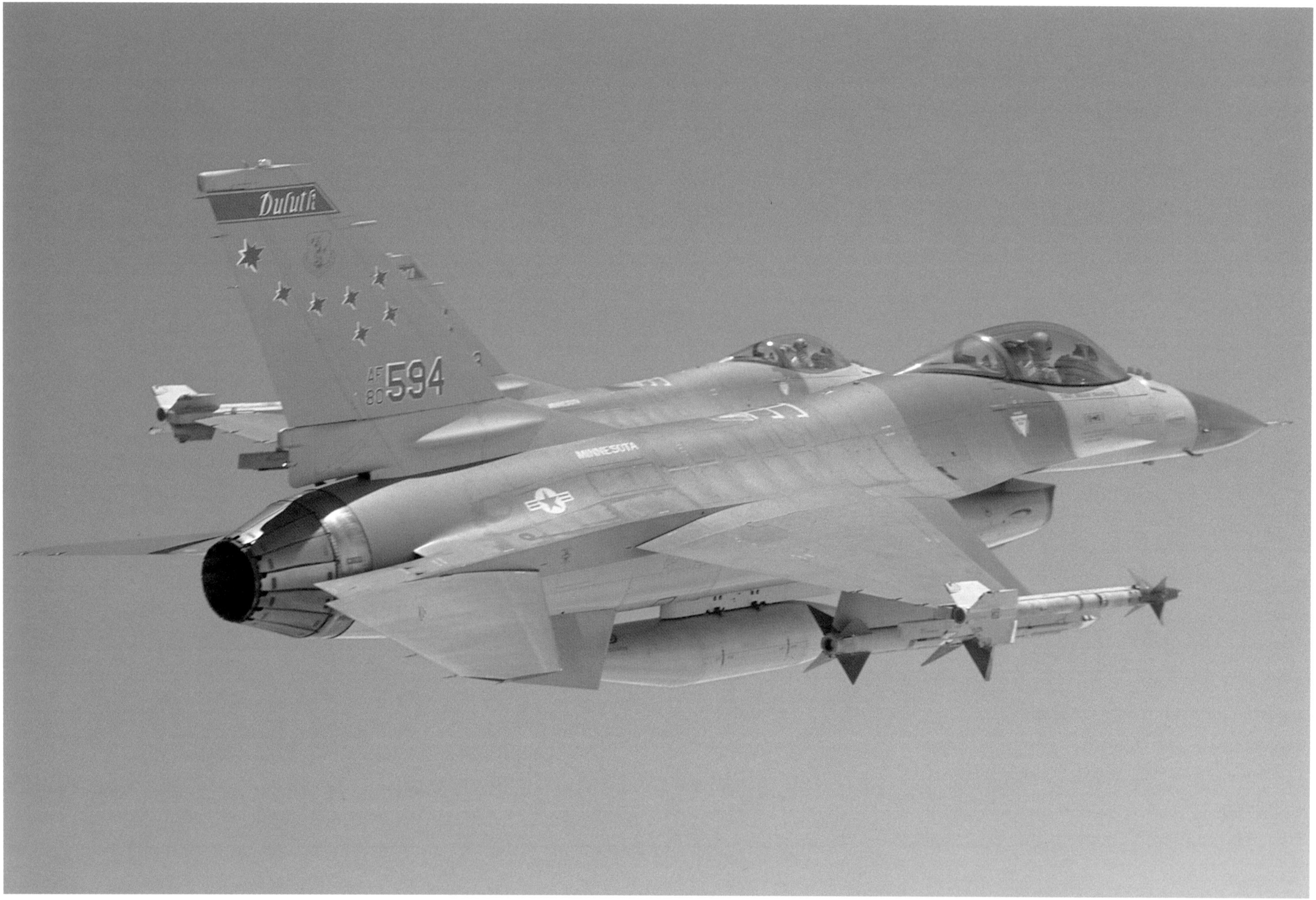

Above Immediately after take-off the Vipers join in close formation and climb to altitude to intercept a non-identified aircraft.

Air National Guard members, alongside Canadian airmen, operate a combined 1st Air Force command post and Continental US NORAD Region (CONR) Air Operations Center to perform the NORAD air sovereignty mission. From its headquarters at Tyndall Air Force Base, Flaorida, 1st Air Force ensures air sovereignty and provides unilateral defense for the United States. First Air Force, comprised of 10 Air National Guard fighter wings and three air defense sectors for the Northeast, Western and Southeast regions of the country, also performs counter-drug surveillance operations in its NORAD role.

No longer armed with muskets on the village green, our modern-day minutemen and women stand ready with sophisticated electronic monitoring equipment and modern fighter aircraft. Our founding fathers could have hardly imagined it. First Air Force fighter wings fly the best of America's fighter inventory, the F-15 Eagle and the F-16 Fighting Falcon.

Three 1st Air Force fighter wings fly the F-15 Eagle: the 102nd Fighter Wing, Massachusetts; the 125th Fighter Wing, Florida; and the 142nd Fighter Wing, Oregon. Seven fighter wings in 1st Air Force fly the F-16 Fighting Falcon: the 119th Fighter Wing, North Dakota; the 120th Fighter Wing, Montana; the 144th Fighter Wing, California; the 147th Fighter Wing, Texas; the 148th Fighter Wing, Minnesota; the 158th Fighter Wing, Vermont; and the 177th Fighter Wing, New Jersey. First Air Force relies upon all of its fighter wings to perform the air sovereignty mission and to be full partners in the Expeditionary Air Force.

This means 1st Air Force pilots are not only masters of air-to-air combat, but of air-to-ground combat as well. As air defenders and partners in the EAF, 1st Air Force fighter wings are prepared to fly and defend America both at home and abroad. The F-16 is the backbone of that defense.

In 1987, the Air Force directed modification of the F-16A for air defense flying. The F-16 Air Defense Fighter employs AIM-7 radar-guided missiles, and is flown by several 1st Air Force units. First Air Force's F-16 Air Defense Fighters have supported the EAF through a multitude of missions, from counterdrug operations in South America to enforcement of the No-fly Zone over Northern Iraq.

The country relies on the Air National Guard in fulfilling these important tasks. Full integration of the Guard into the AEF is a cornerstone of the EAF concept. This realization led 1st Air Force Commander Maj. Gen. Larry K. Arnold to "mainstream" 1st Air Force in late 1998.

The effort to ensure all fighter wings, command and control elements and air battle managers in the organization were trained and equipped to perform the full spectrum of Air Force missions at home and abroad, met with great success.

Now, in the year 2000, all 1st Air Force fighter wings, both dedicated and general purpose, will become integral components of four of the Air Force's 10 AEFs. First Air Force and the Guard are crucial for this concept to succeed. Not only do we provide force protection for America on its own soil and around the globe, we epitomize the Total Force.

The men and women of 1st Air Force have led the way in rediscovering our militia roots. They are entrusted with the all-important Air Force mission of aerospace defense of the homeland.

When historians reflect on the 20th century, they will realize the people of 1st Air Force personified the wisdom of the founding fathers by being true citizen-soldiers with a stake in our own security.

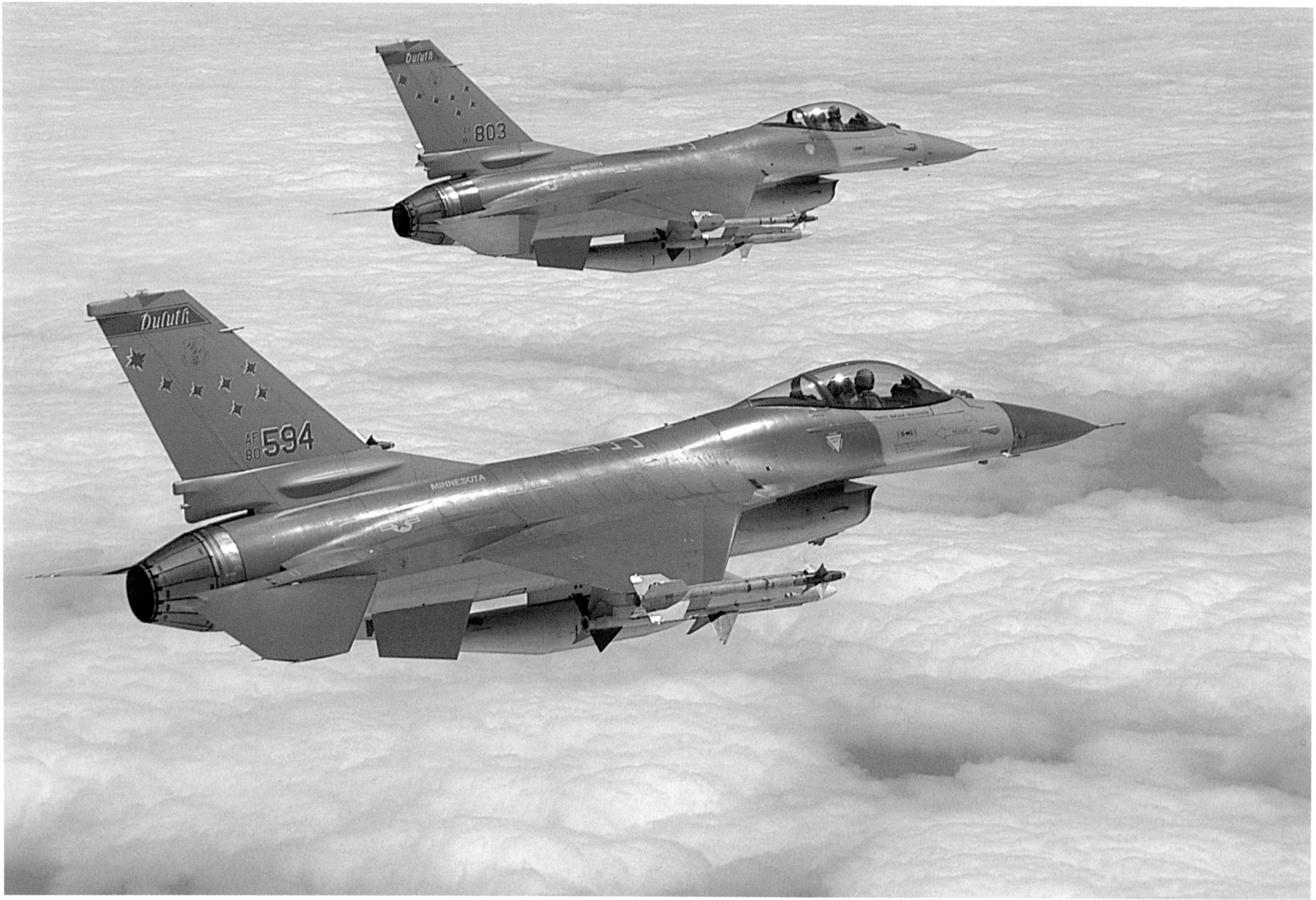

Above and in the page on the right Vipers proceed towards the target assigned by a GCI Center. **Overleaf** Intercept procedures often require the fighters to break formation.

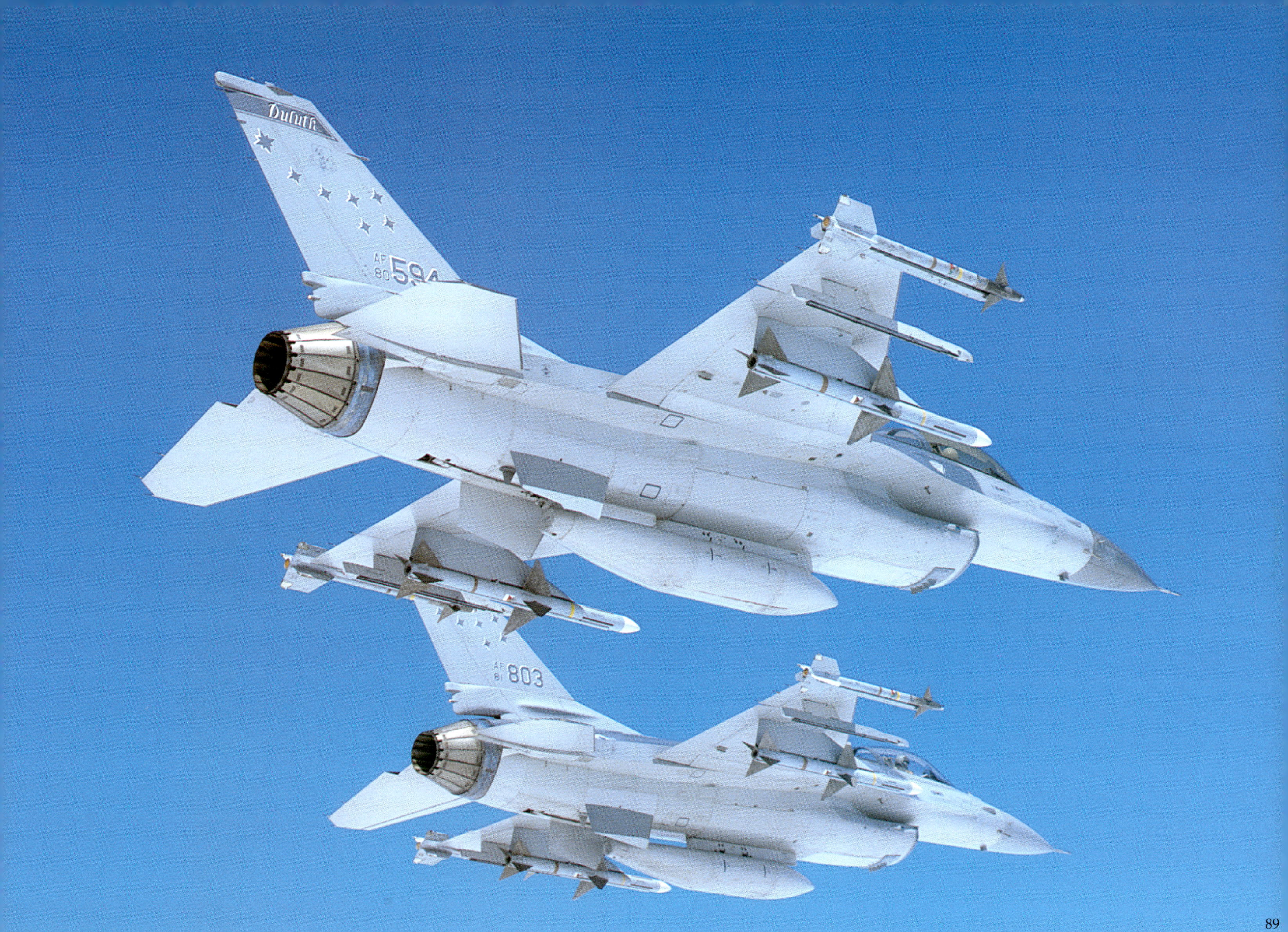
Duluth
AF
80
594
AF
81
803

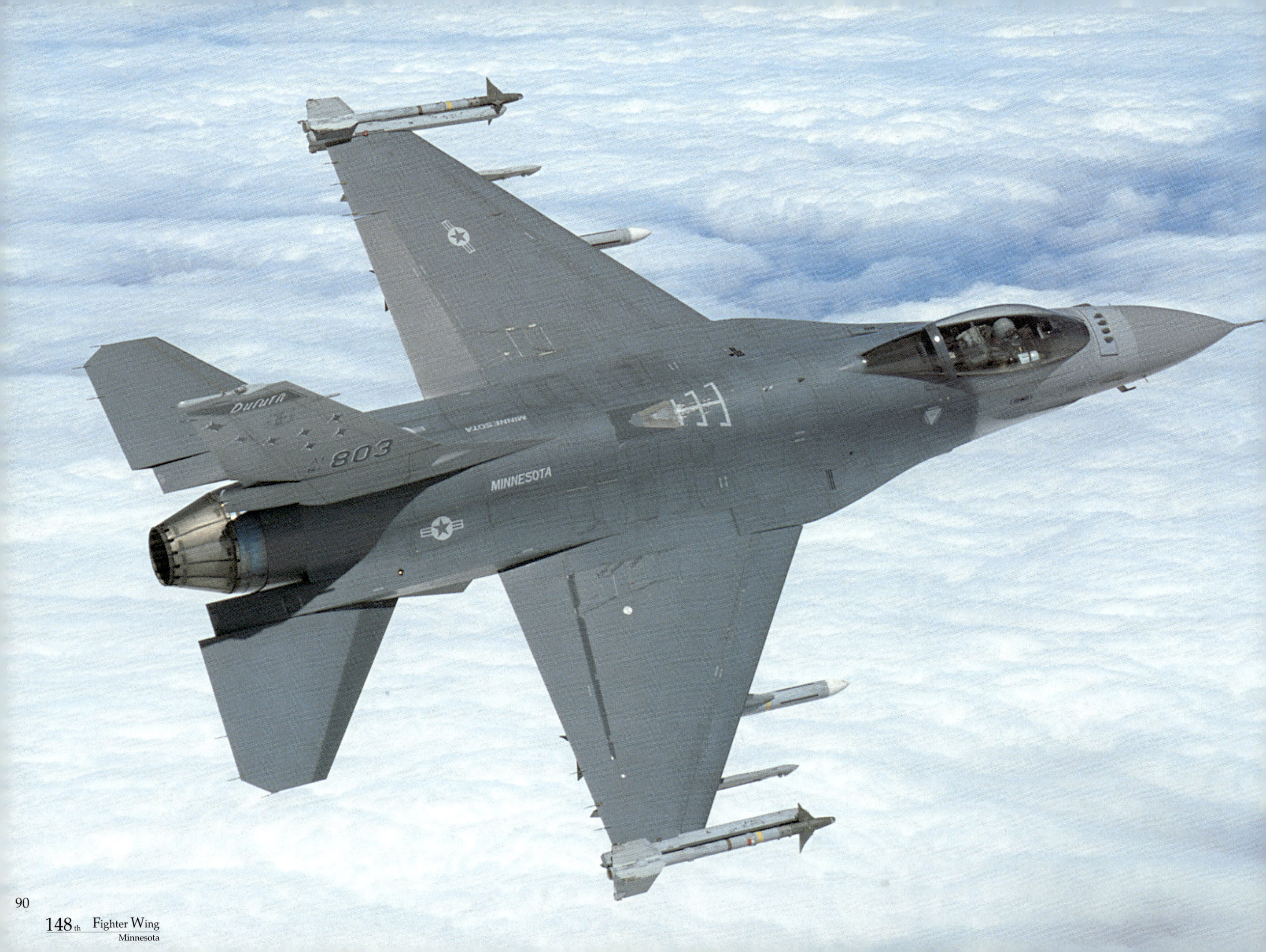
Duluth
803
MINNESOTA
MINNESOTA

150th Fighter Wing

New Mexico

150th Fighter Wing, New Mexico (emblem approved 20 September 1990). The Air Force yellow and blue in the design represent the sun and sky of New Mexico. The yucca plant, river and montains represent the topography of the state while the "Zia" sun symbol represent its native American heritage.
The symbol of flight represents the military preparedness of the unit to defend against and deter all enemies.

188th Fighter Squadron, New Mexico (emblem approved 8 January 1954). The chaparral, or "roadrunner", is renowned for its speed in seeking out ratttle snakes, feinting them into striking and then killing them. This symbolizes the ability of squadron mission. The bird carries two bombs which represent the squadron mission. The bonfire red and Florida gold are the squadron colors.

188th Fighter Squadron, New Mexico "Roadrunners". Constitued at Key, Mississippi on 25 January 1943 as the 507th Bomb Squadron (Dive). Redesignated as the 507th Fighter Squadron and shipped to England in Spring 1944, assigned to the 404th Fighter Group, 9th Air Force, equipped with Republic P-47D Thunderbolts. The unit flew combat missions in the ETO from May 1944 to VE Day. The unit was reconsituted, redesignated the 188th Fighter Squadron (Single Engine) and alloted to Kirtland Air Base, New Mexico, on 7 July 1947. The unit has used five types of aircraft in 44 years: F-51D, F-80C, F-100A/C, A-7D/K, and F-16C/D. In April 1958, the 188th Fighter Interceptor Squadron was the first ANG unit equipped with supersonic fighters and was mobilized for the Pueblo Crisis. It deployed to Tuy Hoa Air Base, Vietnam, on May 1968 for combat operations. The 188th returned to state control on 5 June 1969 at Kirtland AFB. In Fall 1973 it converted to A-7Ds and then to F-16Cs in May 1992.

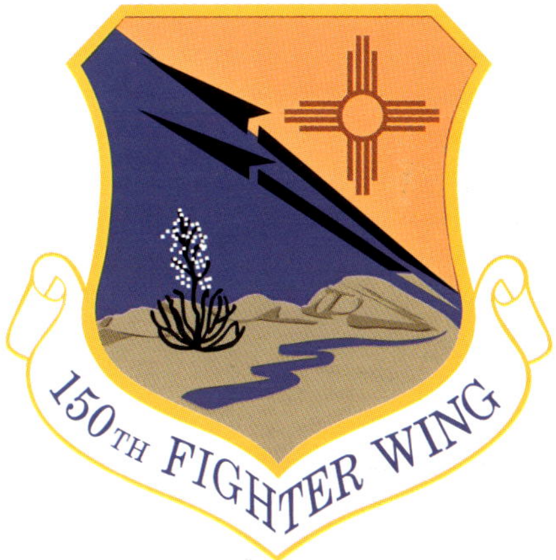

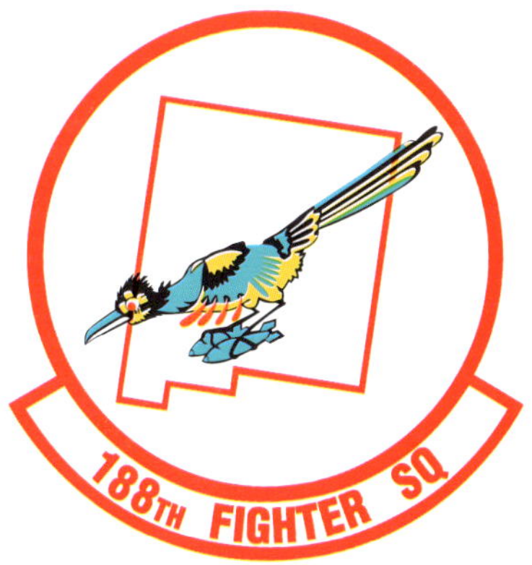

FLYING WITH THE ROADRUNNERS

BY GIAMPAOLO AGOSTINELLI

At Kirtland Air Force Base, New Mexico, an Air National Guard squadron equipped with F-16Cs Block 40 has its nest, and it keeps alway high his "ready for action" status: in fact it is always ready to deploy in a very short time to whatever "hot spots" are in the world, where its action is needed.

The Quick Reaction Force concept requires a very demanding and "quality" training, and to execute many missions that simulate as close as possible those situations which are likely to be encountered in a complex air warfare theatre of operations.

Some time ago, I have had the chance to participate in one of these missions, which was planned to prepare and verify the crews capability to attack an Iraqi air base and destroy its hardened shelters, defence armoured vehicles and aircraft on the ground.

Moreover, to conduct air-to-ground engagements both BVR, Beyond Visual Range, and close-quarter dogfights with adversary's fighters.

It is very important to familiarize with the operational potential of the F-16 Block 40, as it is a real multi-mission-capable combat plane at hand, once its operational characteristics are dominated.

A powerful 29,000-lb General Electric F110-GE-129 augmented turbofan engine and an APG-68 radar with high-resolution ground mapping, and capable to search and track several surface targets both fixed and mobile, make the "Block 40" a first-class fighter.

In the page on the left
A crew chief gives the "stop" signal to a 118th Fighter Squadron F-16C pilot.

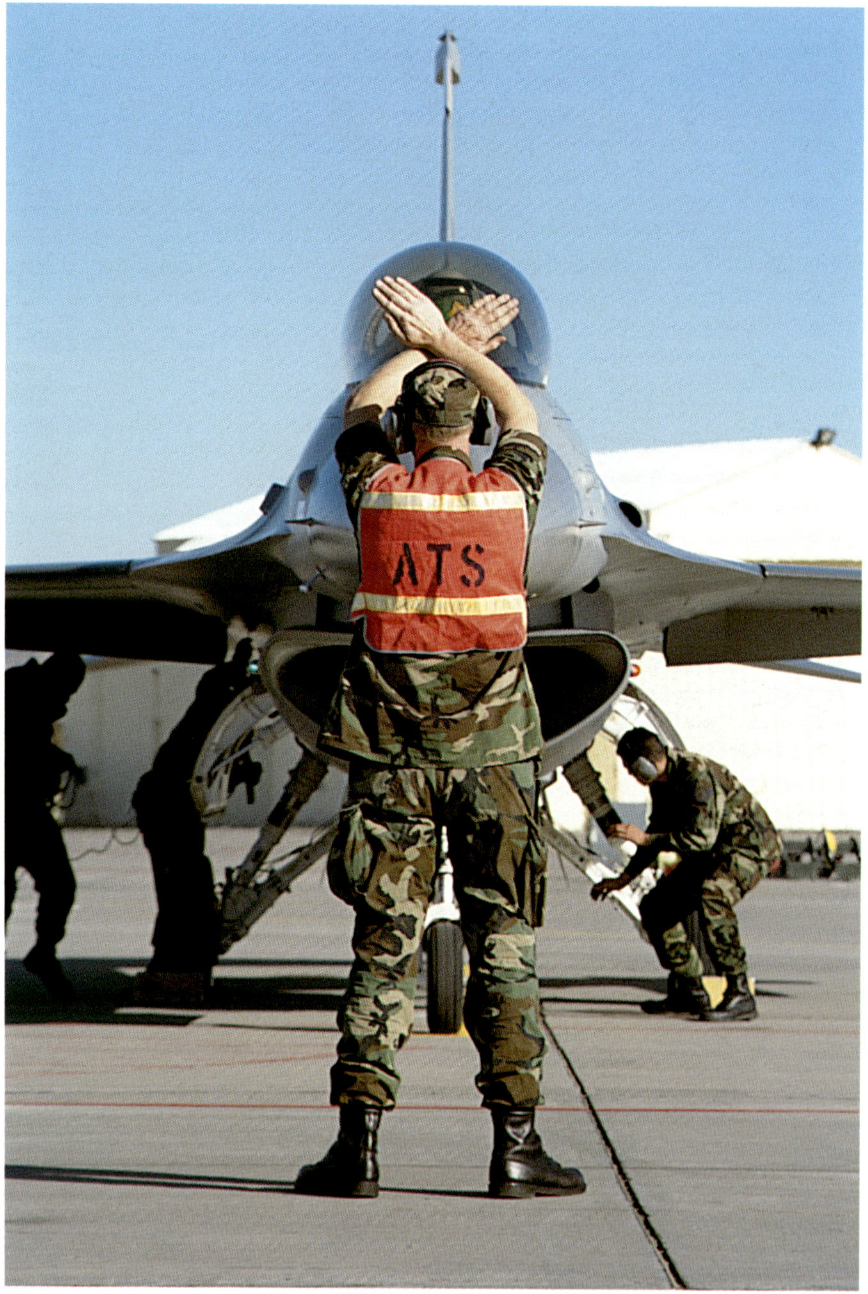

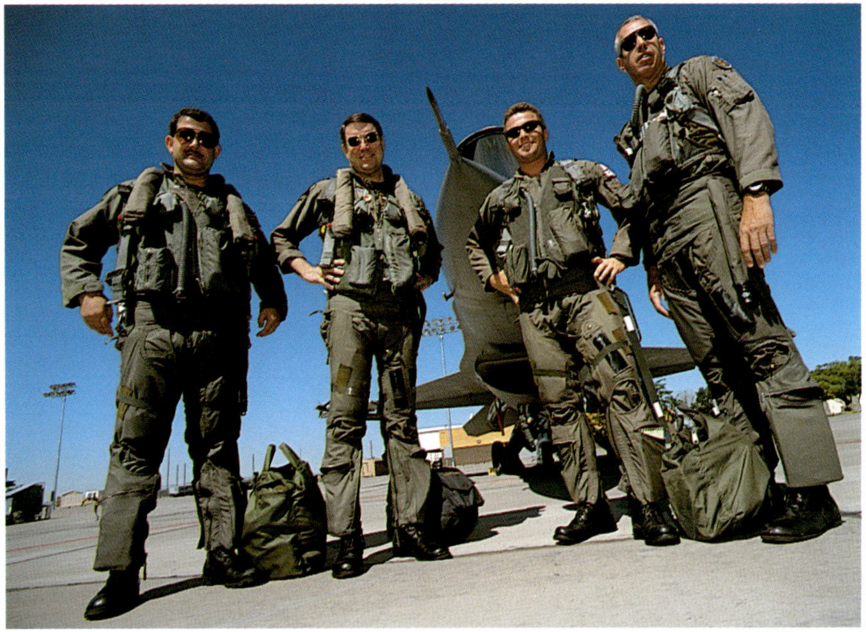

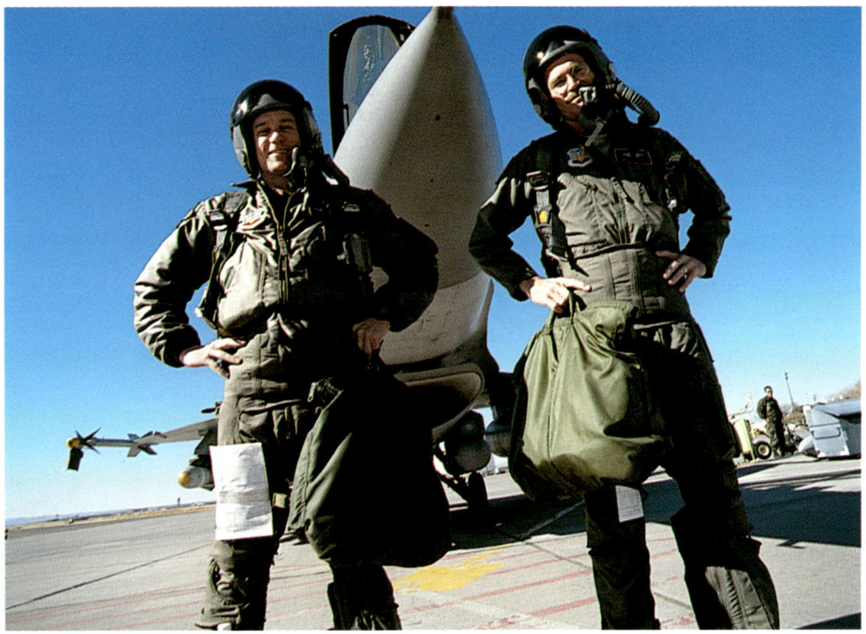

Moreover, the AIM-120 AMRAAM missile gives it the capability to engage several aerial targets beyond visual range (BVR). The sophisticated LANTIRN is also integrated in this advanced version of the Fighting Falcon, giving it the capability to navigate, acquire and designate ground targets at night and to attack them with deadly precision.

Our take-off is planned at 09:00, but we meet in the briefing room two hours earlier to prepare and verify the entire mission in the finest details, because it is going to be particularly complex and difficult. We will fly with Lt.Col. Ray Mahalick, Maj. Mark McCausland and Capt. Tony Levin. Maj. John Heard will fly the F-16D in which I am going to be the back seater. John has been a Navy pilot for several years and he has logged some 1,000 hours in the F/A-18 before joining the Air National Guard. He is a good and experienced pilot, but he is not the only one in the 150th Fighter Wing.

The Air National Guard counts many ex-USAF and some ex-US Navy and ex-US Marine Corps pilots in its ranks. This helps keeping a higher readiness and operational capability in its units, making them available for real operations that once were exclusive jobs for the "active duty" Air Force.

Lt.Col. Mahalick starts the briefing illustrating the main mission objectives: radar procedures rehearsal and simulated launches of AIM-120 AMRAAMs in BVR combat situation, then use of the LANTIRN system with simulated launches of AGM-65D Maverick IR-guided air-to-ground missiles, ending with gun and free-fall practice bomb attacks on a range south-west of Albuquerque.

Before we can get to our target area, our flight will be attacked by two other

On the left
Posing after a mission
on the 150th FW flight line
at Kirtland AFB.

On top
Four 188th FS pilots ready
for their training sortie at
the firing/bombing range.

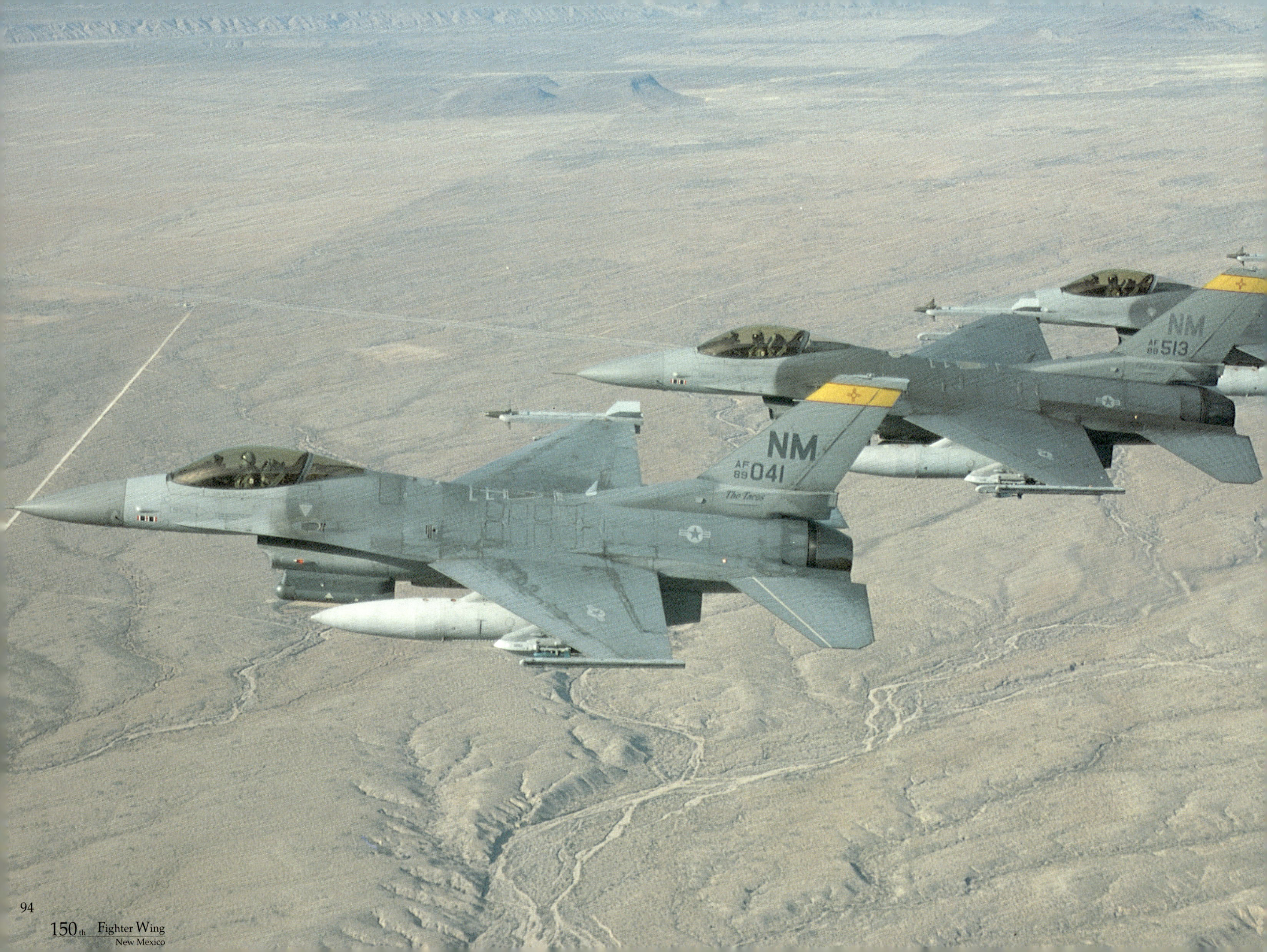
NM
AF 89 041
NM
AF 88 513

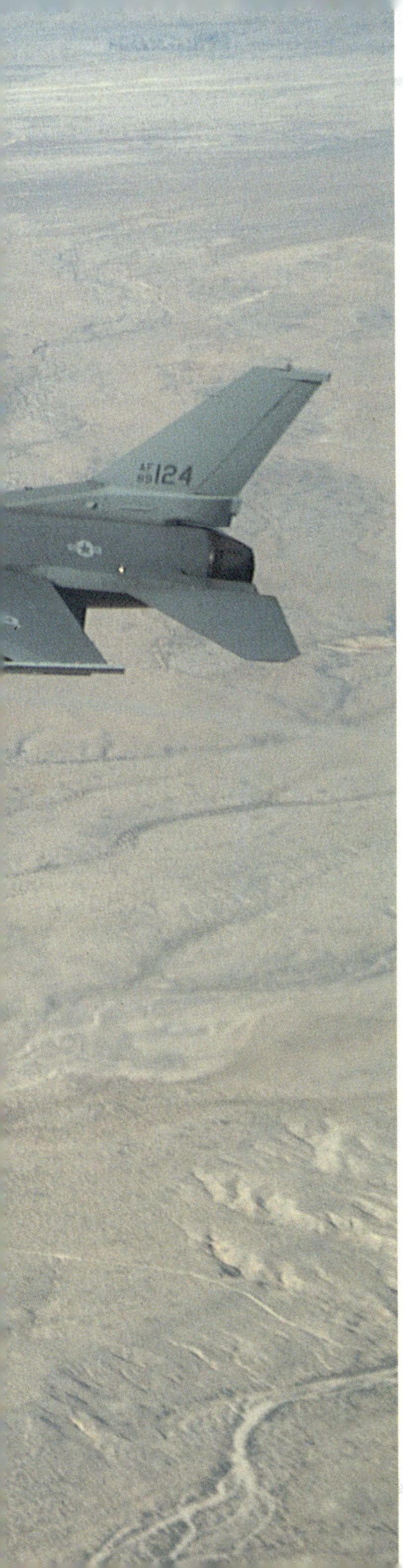

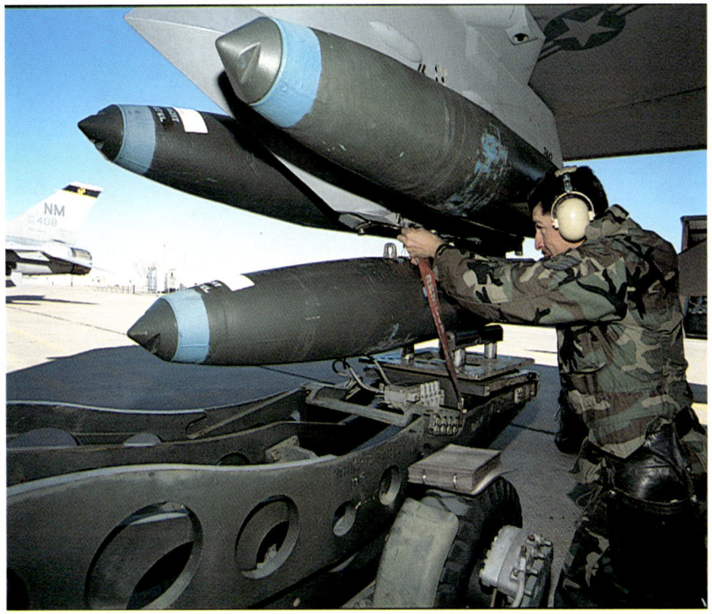

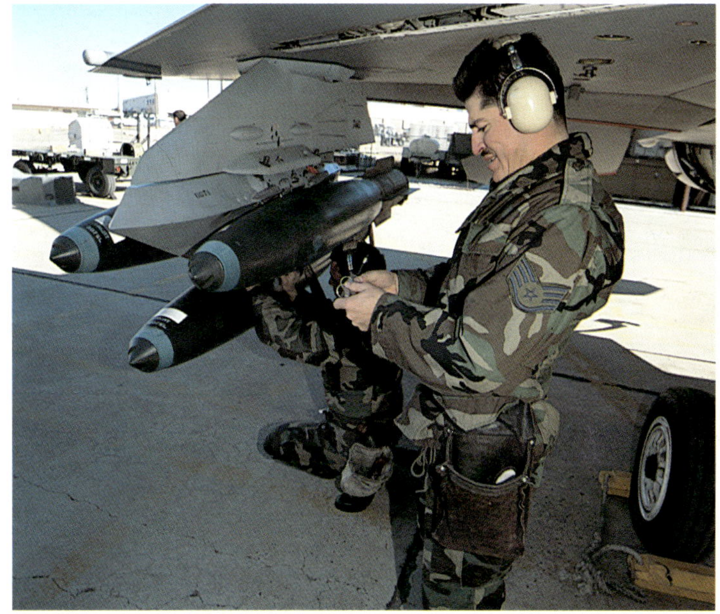

F-16s simulating Iraqi MiG-29s. Their pilots joined us briefly in the briefing room. Air-to-air engagement rules were discussed: levels, speeds, disengagements. They are to reproduce the Fucrum operational characteristics in combat as close as possible, including tactics and maneuvers utilized by Iraqi pilots from the Gulf War onwards. Latest samples were recorded in December, 1999.
Times between turn points in the low-level flight which will lead us South almost until the White Sands desert are discussed. Thence we will turn to north-west to reach our targets: battle tanks, a truck convoy and some Anti-Aircraft Artillery (triple A) sites.
When the briefing is over, Major Heard takes me "on charge" to discuss the latest details and all the emergency situations that could arise from start-up to our return to the blocks on the tarmac.
At 08:10, flight suits checked, we step to the aircraft assigned to us on the flight line. While John is busy with the external ground check or our twin seater, I strap the rear seat's Boeing ACES II belts on,

Above and on the left
Weapons specialists train often with the many various F-16C Block 40 weapons. The ones seen in these pictures are normal free-fall prectice bombs.

In the page on the left
A tight echelon formation of three 150th FW F-16C Block 40 Fighting Falcons approaching a firing and bombing range in the New Mexico desert.

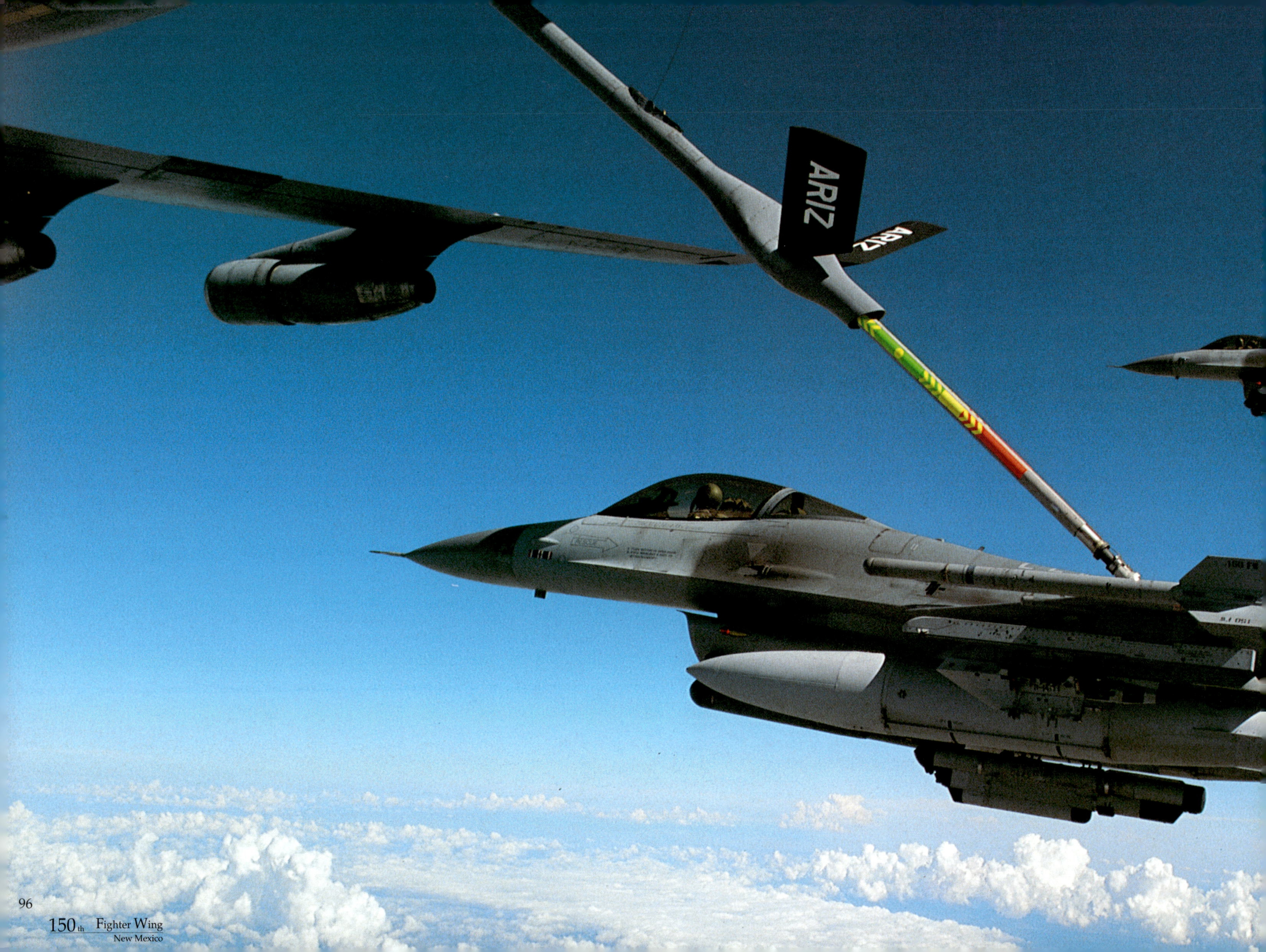
ARIZ
ARIZ

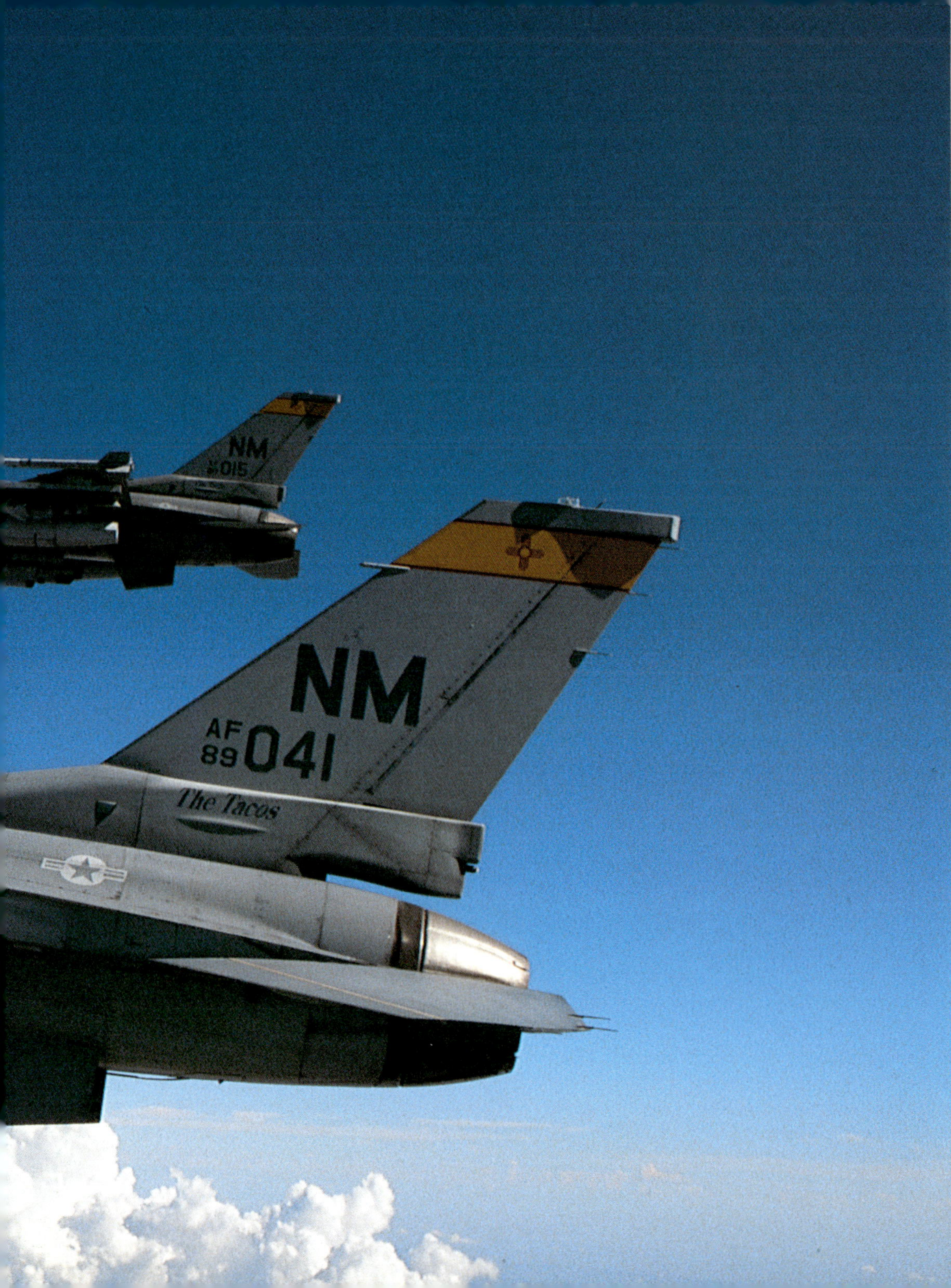

Then I plug my oxygen mask's tube in, check my photo gear for the last time, and wait for John's turn. After he jumped in the front seat and strapped himself, the General Electric F100 start-up sequence is initiated. It's 08:39.

A low-pitch hiss is barely audible in spite of the rpm revving up, as soon Major Heard closes the long and wide transparent canopy.

"Some more time, to allow averyone to be ready for taxiing. In the mean time we 'hone' the targeting pod, which you can see at work in your right-hand multi-function CRT. Now, look how the FLIR turret is turning" John advises.

I look with curiosity and wonder the characteristic infrared image while the rotating head of the AN/AAQ-14 pod moves every 3-4 seconds with brief, sudden displacements until stabilizes itself onto the hangar in front of us.

We taxi forming a column until the North runway head, following the other three Fighting Falcons.

After our weapons safety pins are removed by the specialists, at a dedicated pad, we are ready to take-off in two pairs, ourselves being in the second row. We will rejoin at 6,000 feet, heading 190°, then we will climb to 16,000 ft, descending to low level in the proximity of the target area in order to mask ourselves behind the terrain

On the left
Two 150th F-16Cs refuel at an Arizona ANG KC-135.

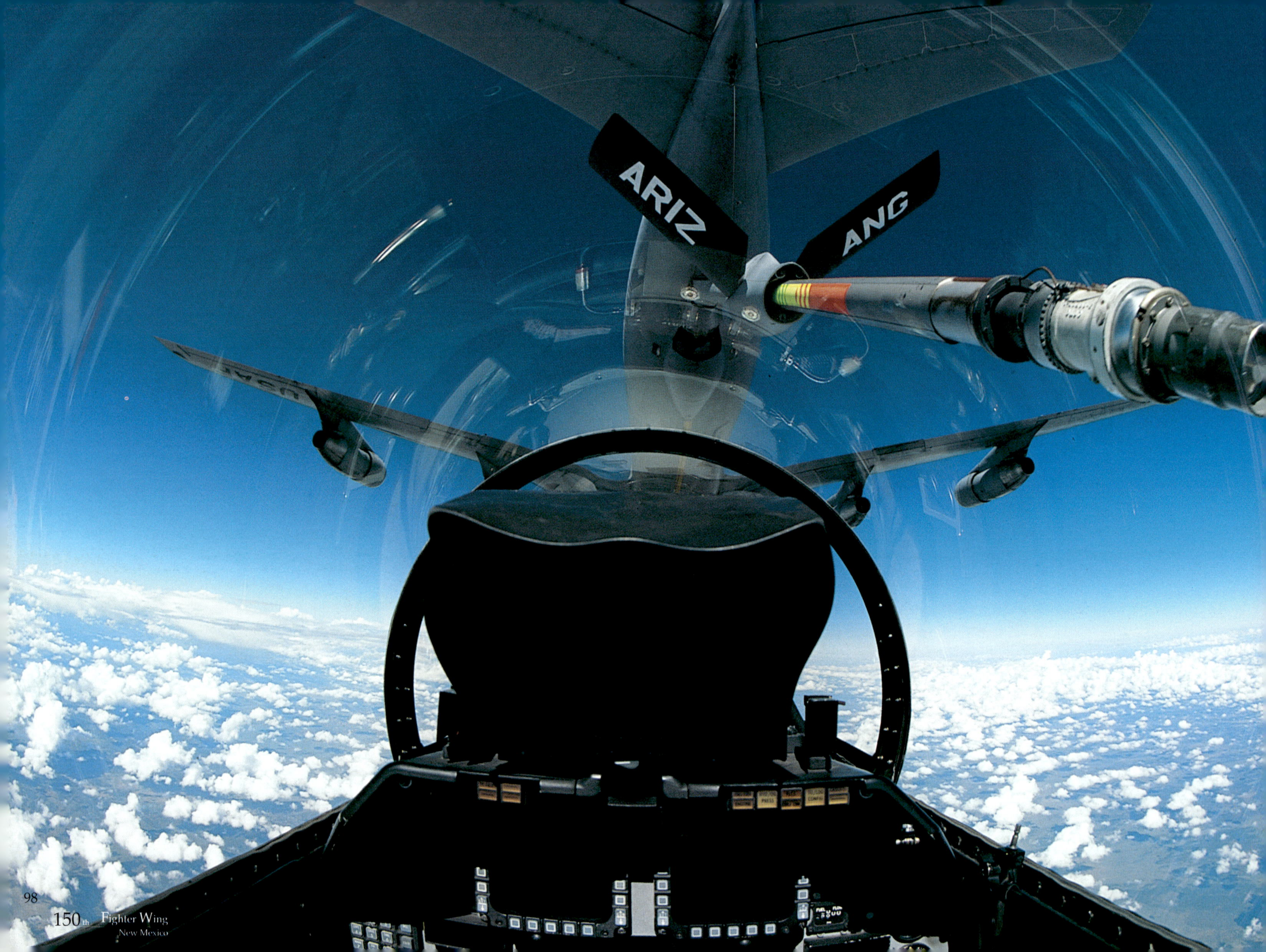
ARIZ
ANG

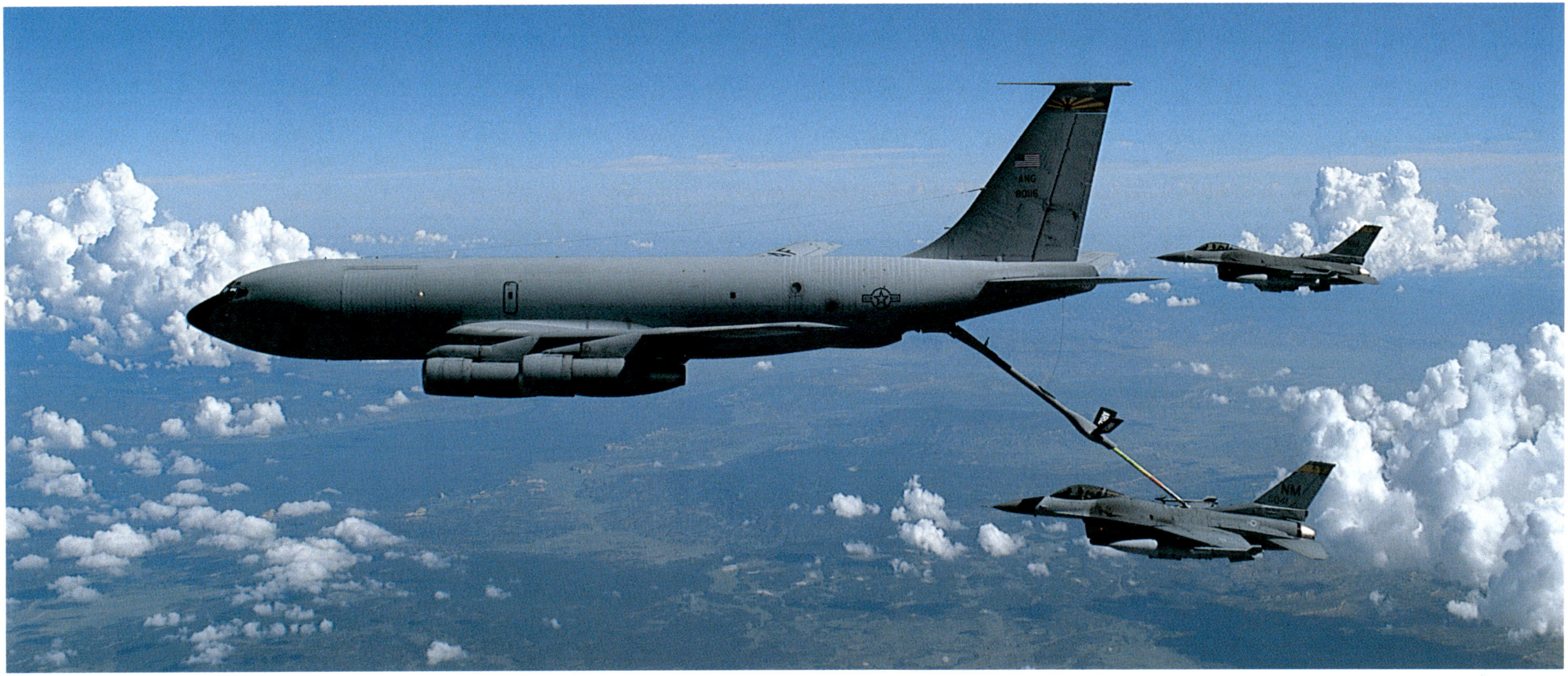

before climbing again to no less than 10,000 ft in the final run-in to our targets.

This because the recent wars and International crisis - Kosovo being the latest example - taught that attacking a target using "smart" weaponry flying at low level is useless because it makes more likely being hit by all that hardly spottable and deadly AAA.

The first pair is roaring down the runway: 5, 6, 7...8, go! John releases the brakes and some instants later, typical of jet planes, I feel the powerful thrust of our afterburner, like a mighty hand pushing us down the runway.

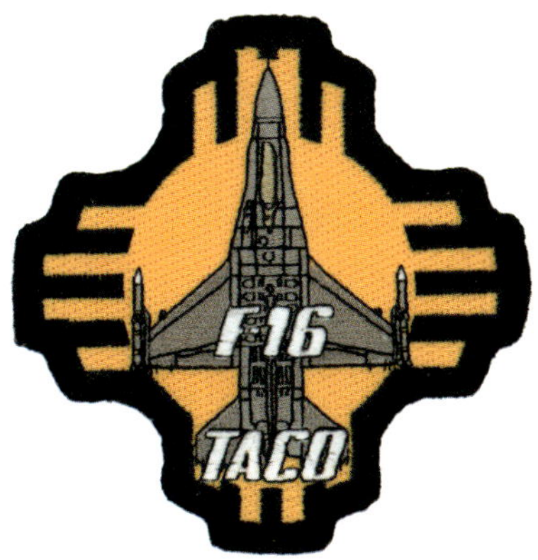

Above
A 150th Fighter Wing F-16C is being refueled while a second one is waiting its turn, off the trailing wake.

On the left
A 150th Fighter Wing Tactical Officer shoulder patch.

In the page on the left
The KC-135's flight refuelling boom as it is seen from the rear seat of a 188th Fighter Squadron F-16D.

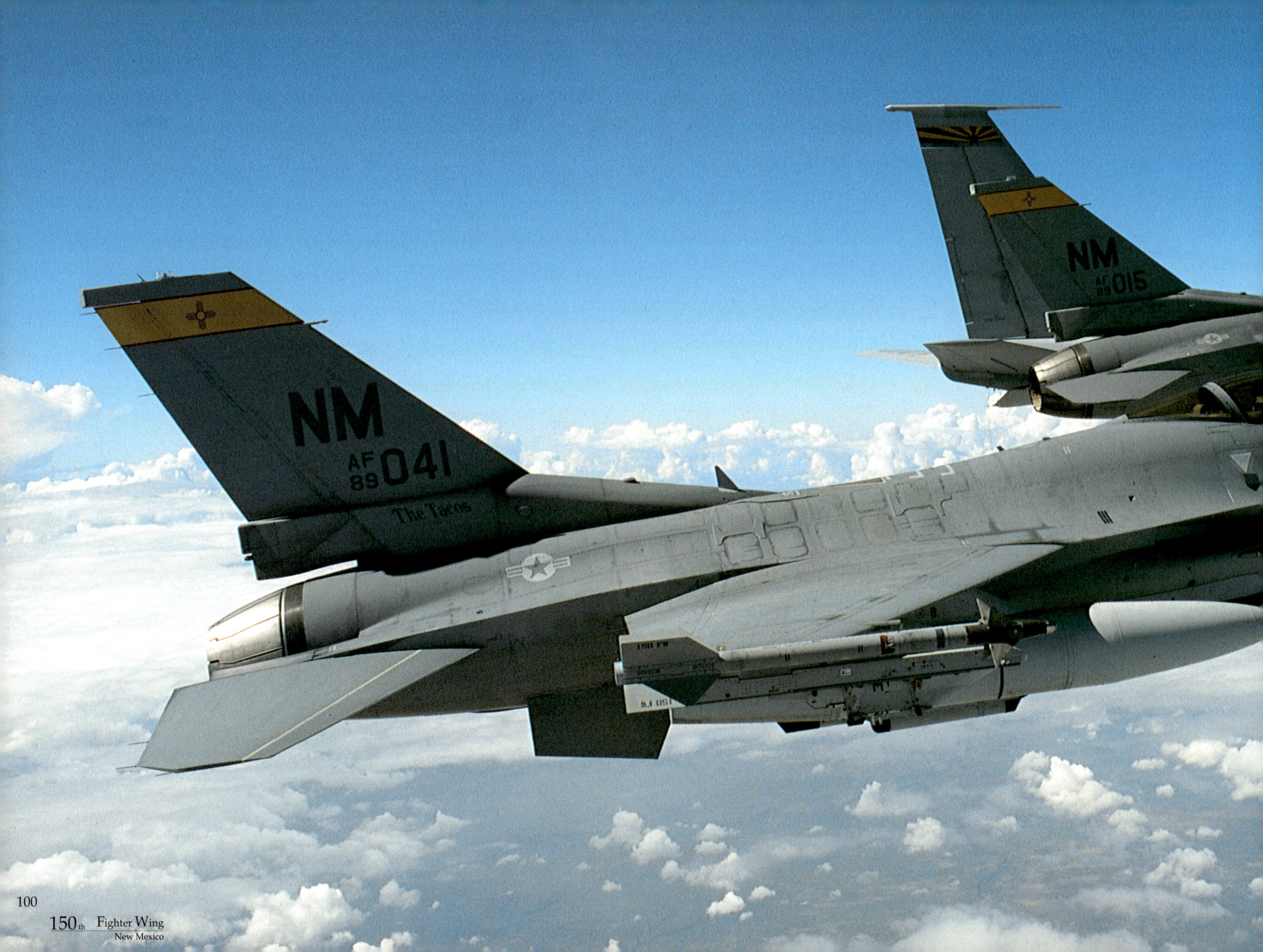
NM
AF 89 041
The Tacos
NM
AF 89 015

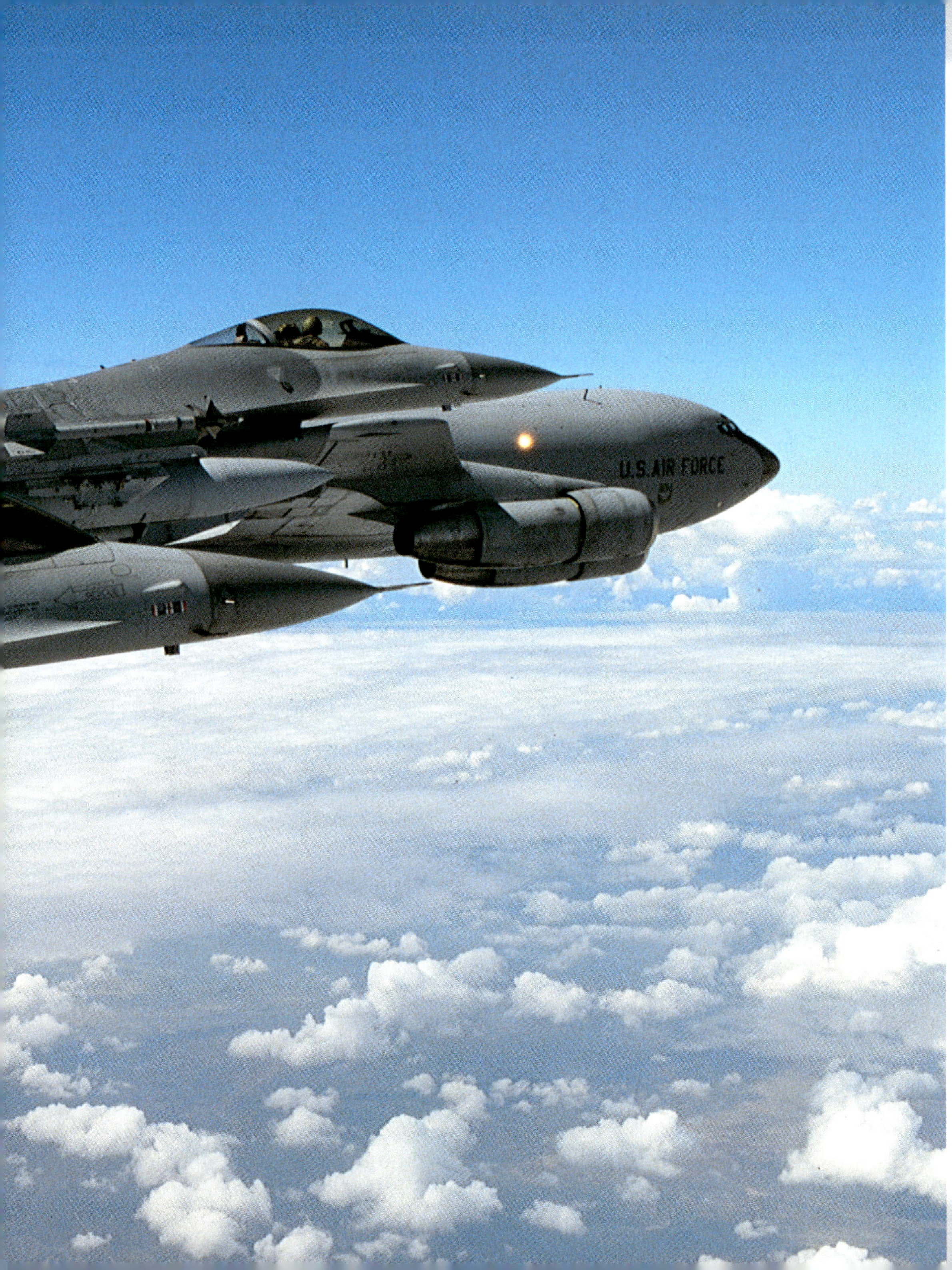

Our wingman follows us like a shadow and I see it aloft at our side after a mere 800 meters ground run, its wheels retracting in the fuselage belly. Indicated airspeed increases rapidly: 230...250 knots, and also the thrust and fuel consumption: the fuel flow instrument, on the top right panel, shows some 52,000 lb/hr.

Afterburner off, one feels something like braking, but the aggressive Fighting Falcon continues climbing steadily under the 97 per cent thrust set by John. With a perfect rejoin, we set in a "fingers" formation heading to the first turn point.

"Here we are, now we descend for the low-level portion" John explains. At a Lt.Col. Mahalick order, we part a while to descend, and then fly some 80 meters above the desert floor. 15 minutes to the next turn point, a bridge over a dry stream.

Thereafter we will climb again to 12,000 feet. "Right, now we climb, heading two-four-zero" John advises on the intercom. "OK, now we part in two pairs, so that we will approach the convoy in a way to have our sides under the other pair's radar coverage in case we are attacked by enemy fighters" he adds. "In real operations we would be under AWACS coverage, but we are to train for any eventuality and become expert with all systems available."

On the left
Once filled-up with fuel, the two F-16Cs depart from the tanker.

Our wingman flies parallel to us, some 500 feet apart. Our two AN-APG-68Vs - the F-16 Block 40 radar - are scanning in search of possible enemy fighters. On my head-down display on the right I can see the scan: nothing in view. We exchange data to compare the tactical information.

"Bandits at 9 o'clock, 20 miles, 14 angels" Mahalick advises from the other pair. "You cover the easterly sector, this one is ours." John turns sharply to the right with a 60-degree bank, and fingering a roller on his throttle lever he tilts the versatile radar scanner upwards.

"I've got the other, you see it?

"You can see it on the right high sector on your screen, at two o'clock... Now by means of another roller, upper on the lever, I move a cursor on that blip to lock it on...

"Weapons panel on, AIM-120s selected. In case of a real foe, that would be confirmed with a threat symbol displayed by the IFF processor.

"Now the cursor is on the target. Lock-on! First AMRAAM, fire!... and now the second, fire!

"We do not have a special pod (*such as the one used in the ACMI, Air Combat Maneuver Installations, ed's note*) that would process our data and the whole situation, therefore confirming the 'real' kills exactly, but we can manage anyway by means of our onboard computer, which ensures a good deal of veracity" Major Heard explains.

The launch data, the flying characteristics of the other F-16 which "acts" as a MiG-29 and the AMRAAM configuration tell us that we have a confirmed kill for the first launch. A few moments earlier, also Lt. Col. Mahalick confirmed his own kill as well. Now we can concentrate on the ground targets. Radar in long-range mode at 10,000 ft: the targeting pod is working and shows us a column of six tanks. "The last in line is mine" I hear John telling to our wingman Captain Lewin. "Weapons panel set on AGM-65D missiles.

"Now I move the cursor on the last tank in the column to lock it on" he explains. The AN/AAQ-14 with its FLIR turret passes all the target data to the IR Maverick head. "Here we are; now the targeting pod will operate automatically and will fire the Maverick at the right moment against the selected tank on which it has been locked. With the same verification method, we have a confirmed kill also of the ground target. I can appreciate the high level or realism of the firing range.

In the page on the right
The New Mexico desert is a natural environment for the 150th FW F-16s.

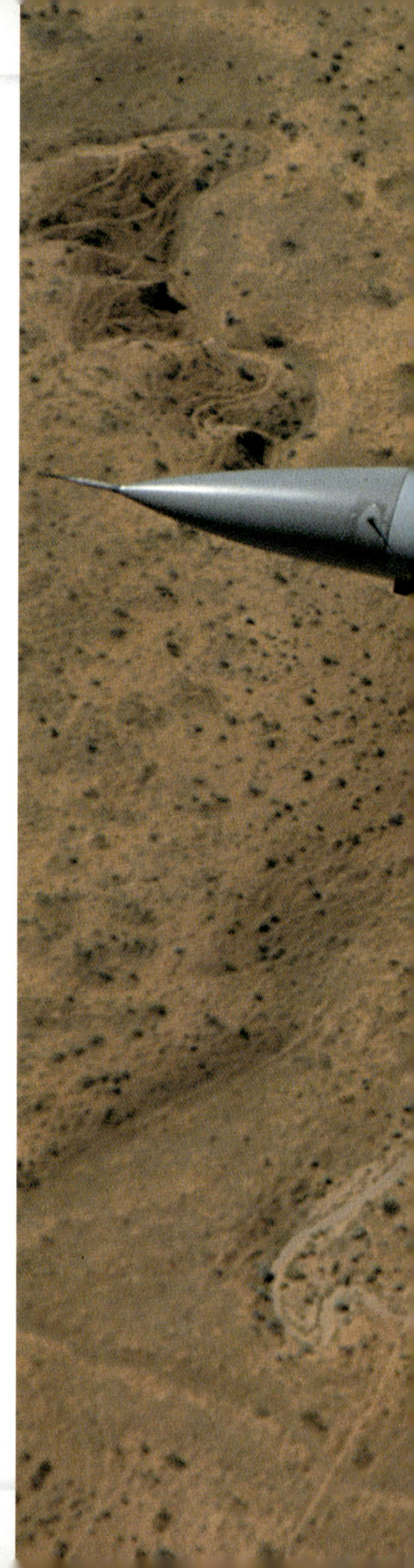

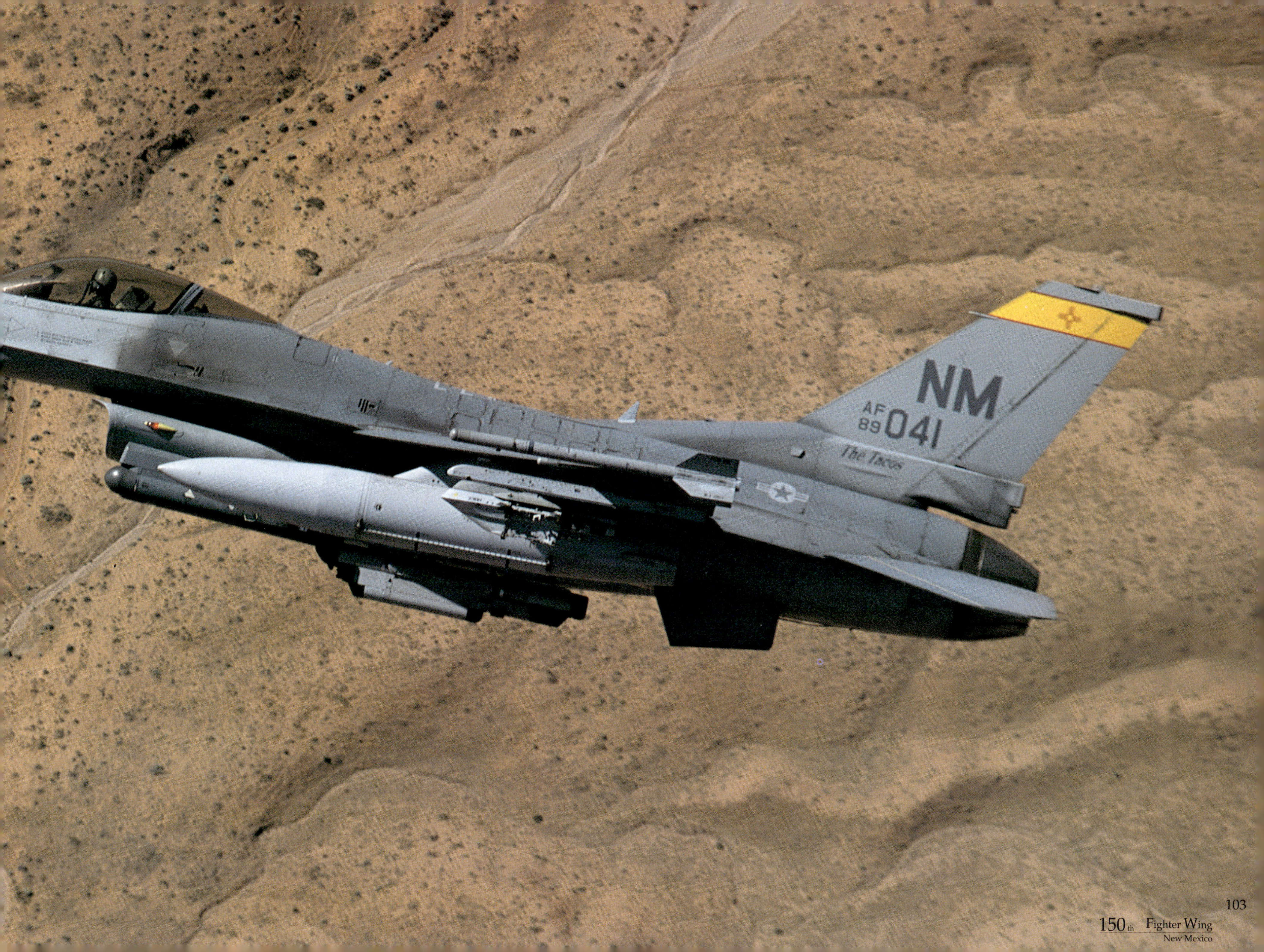
NM
AF
89 041
The Tacos

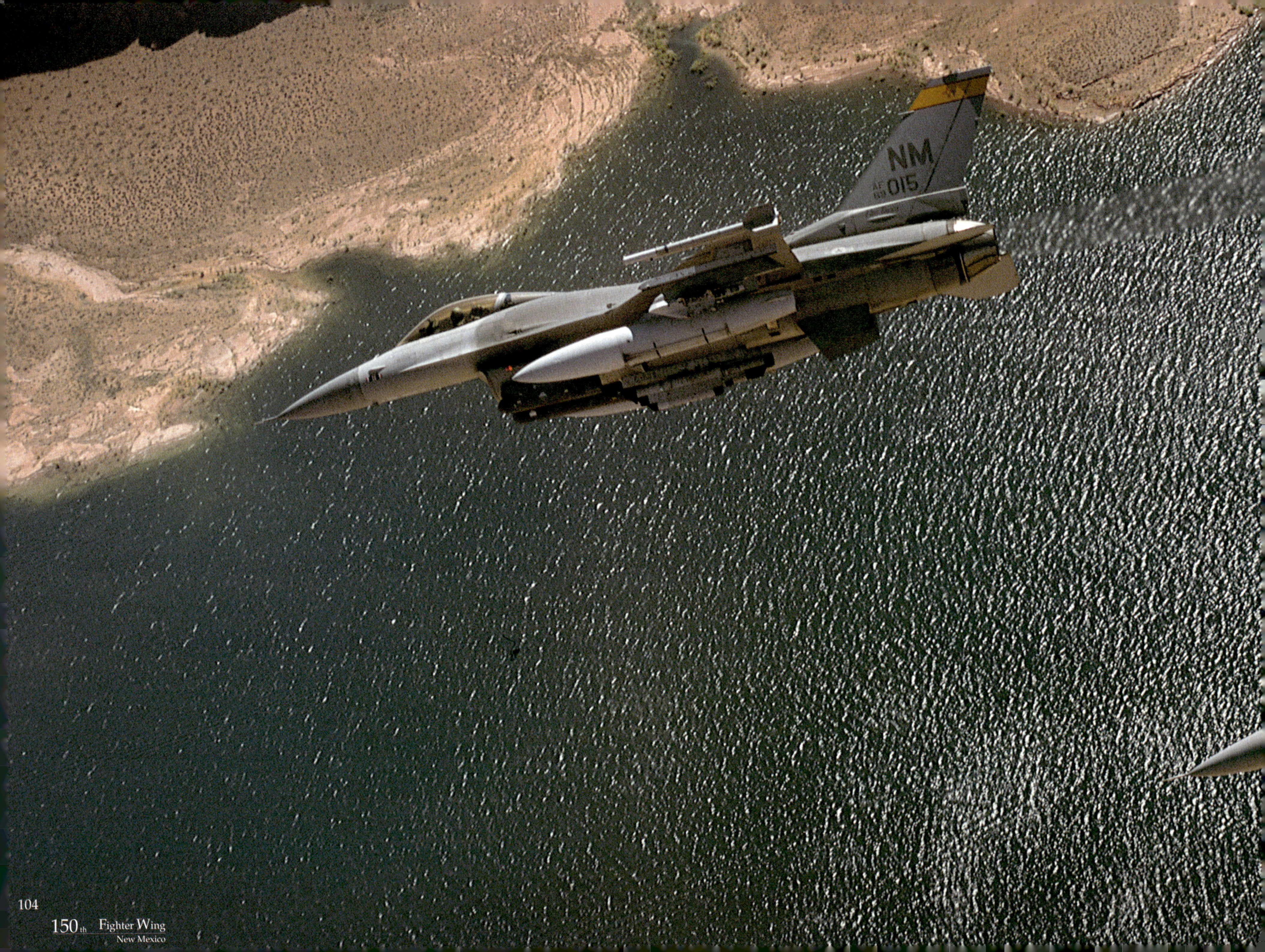
NM
AF
88 015

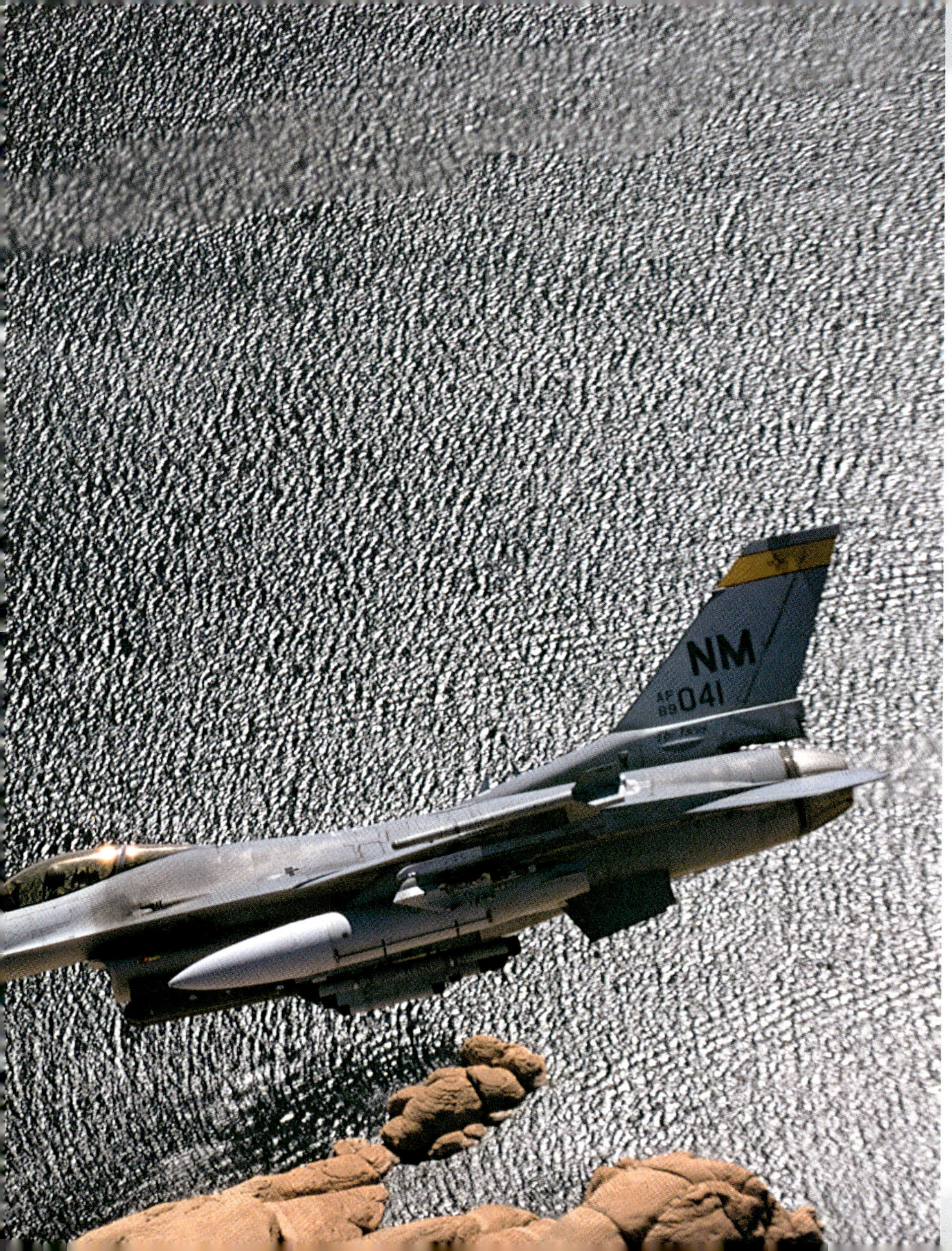

"Now we go with the gun" he advises. We enter the firing range from North, then we climb for a while in a parabolic trajectory which allows us to spot our target better during the wingover maneuver.

"Ready? Here: three, two, one, go!" A sudden noise, vibration and a puff of smoke on my right tell me that our M-61 six-barrel Gatling gun has fired a deadly ripple. John pulls 5.5 Gs at the end of the dive, with our engine in "full military" and immediately starts a tight turn to the left followed by a right one: typical evasive maneuvers.

Then we show up again for another gun pass. Now it's the turn for the practice bombs: definitely a wearying mission. After countless pull-ups and fatiguing positive Gs (and some negative), I feel exhausted, but absolutely full of enthusiasm.

We head home in loose formation and touch down on the runway exactly one hour and forty-five minutes after we left it, with just over 800 pounds of fuel, out of the 10,900 that we had in our tanks on take-off.

Unfortunately it is time to part from this wonderful thoroughbred, but the memory of that experience will live on forever in my mind.

On the left
Overflying the last mission turn point, over the Western coast of Lake Powell.

In the page overleaf
Two 150th FW Vipers overfly the scenic Monument Valley.

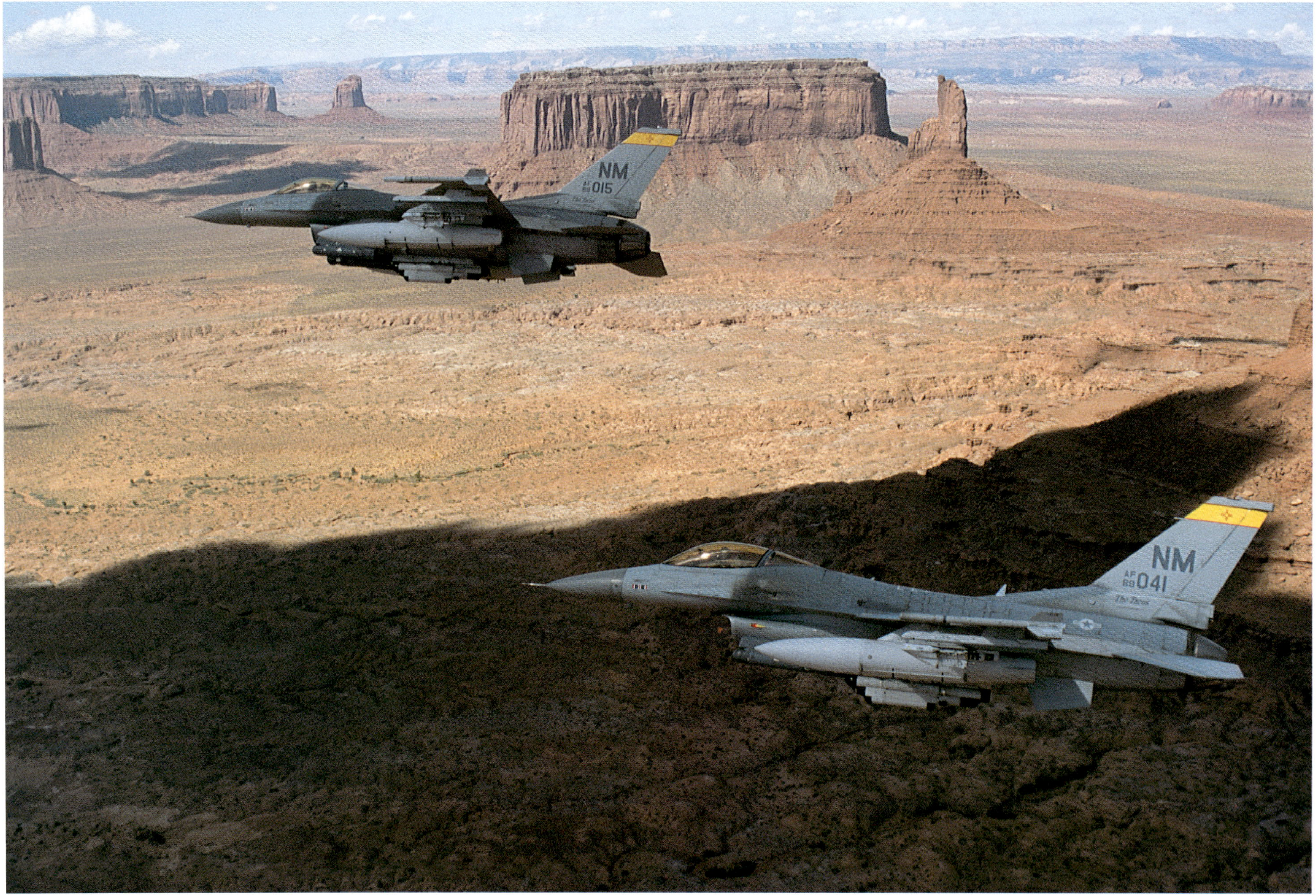
NM
AF 89 015
NM
AF 89 041

169th Fighter Wing

South Carolina

169th Fighter Wing, South Carolina (emblem approved 5 August 1966). The Air Force blue stands for the sky and the red color denotes courage and valor in combat.

The star refers to the unit's participation in the combat theater and campaign credits of the fighter squadron, symbolized by the fox. The Palmetto tree represents the State of South Carolina and its military heritage since 1775. The placing of the stars and Palmetto branches and trunk indicate the unit's number 1-6-9.

157th Fighter Squadron, South Carolina (emblem approved 22 March 1984). The blue represents the sky. Yellow refers to the sun and the excellence of the squadron's personnel. The fox reprents the "Swamp Fox", a hero of the Revolutionary War whose area of operations was in South Carolina. The "Swamp Fox" represents military excellence, cunning, and fearlessness in the face of any enemy or threat.

157th Fighter Sqn., South Carolina "Swamp Fox". Activated at Mitchell Field, New York on 1 October 1942 as the 350th Fighter Squadron flew combat missions in the ETO from Agust 1943 with P-47D Thunderbolts and from October 1944 with P-51D Mustangs until VE Day. Redesignated as the 157th Fighter Sqn. and assigned to the South Carolina National Guard on 24 May 1946. Equipped with P-51Ds at Congaree Air Base on 9 December 1946, the 157th FS (SE) was mobilized for Korean War on 10 October 1950. Since that time, the squadron has been equipped with F-51Hs, F-80Cs, F-86Ls, F-104A Starfighters (with which it deployed to Spain during the Berlin Crisis), F-102As, A-7Ds. Having won the Gunsmoke 1989 fighter competition six years after converting to the F-16A Fighting Falcon, the unit was deployed to Saudi Arabia for Operation *Desert Storm*. The unit was mobilized on 29 December 1990 and deployed to Al Kharj Air Base where, as part of the 4th Tactical Fighter Wing (Provisional), it flew 1,729 sorties without a loss. It converted to F-16C/Ds in 1994.

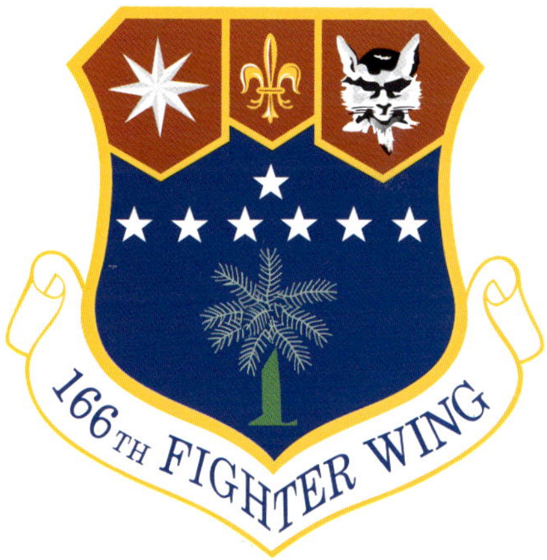

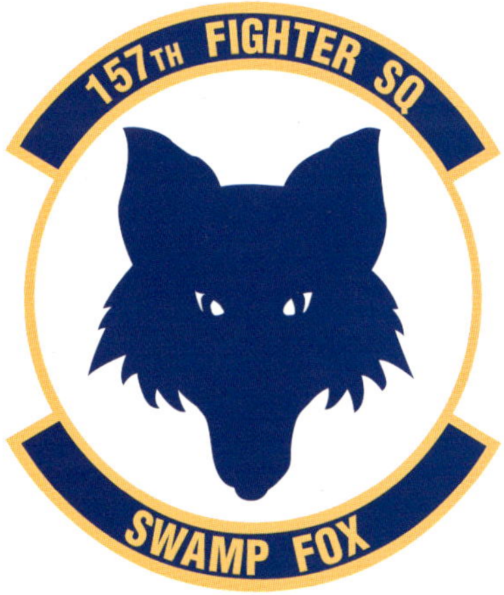

SHIELD TO STORM

BY CHARLES J. GROSS

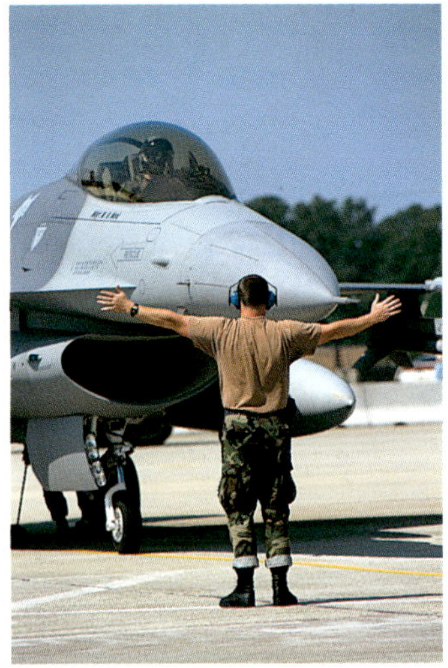

Air Guard fighters merged well with active duty squadrons in the 4th Tactical Fighter Wing (Provisional) at Al Kharj. They participated in the air campaign from the first day. By the time the war ended, the ANG's F-16s had flown 3,645 missions and dropped 3,500 tons of ordnance without losing a single aircraft to enemy fire or accident. Although the New Yorkers flew close air support missions, they and the South Carolinians adapted quickly to the demands of Persian Gulf operations. Just two weeks after deploying to Saudi Arabia, both units found themselves in combat. Lacking the equipment needed for night and weather operations, they staged a variety of daylight attacks on targets in Kuwait and Iraq.

Those included SA-3 missiles sites, airfieds, bridges, and the elite Republican Guard. Their targets were located in Kuwait and Southern Iraq. Although electronic countermeasures and Wild Weasel support were not directly attached to ANG F-16 formations, for istance, South Carolina's first mission was to suppress enemy air defense in support of a strike package tasked to bomb Ahmed Al Jabar and Ali Al Salem Airfields in Kuwait.

Consequently, the squadron attacked 10 Iraqi SAM sites in marginal weather conditions. During that mission they "... soaked up more SAM's than they care to see again. SAM 'wiggles', chaff pods, RWR... and eyeballs (as well as good prayers from the Chaplain) got them through."

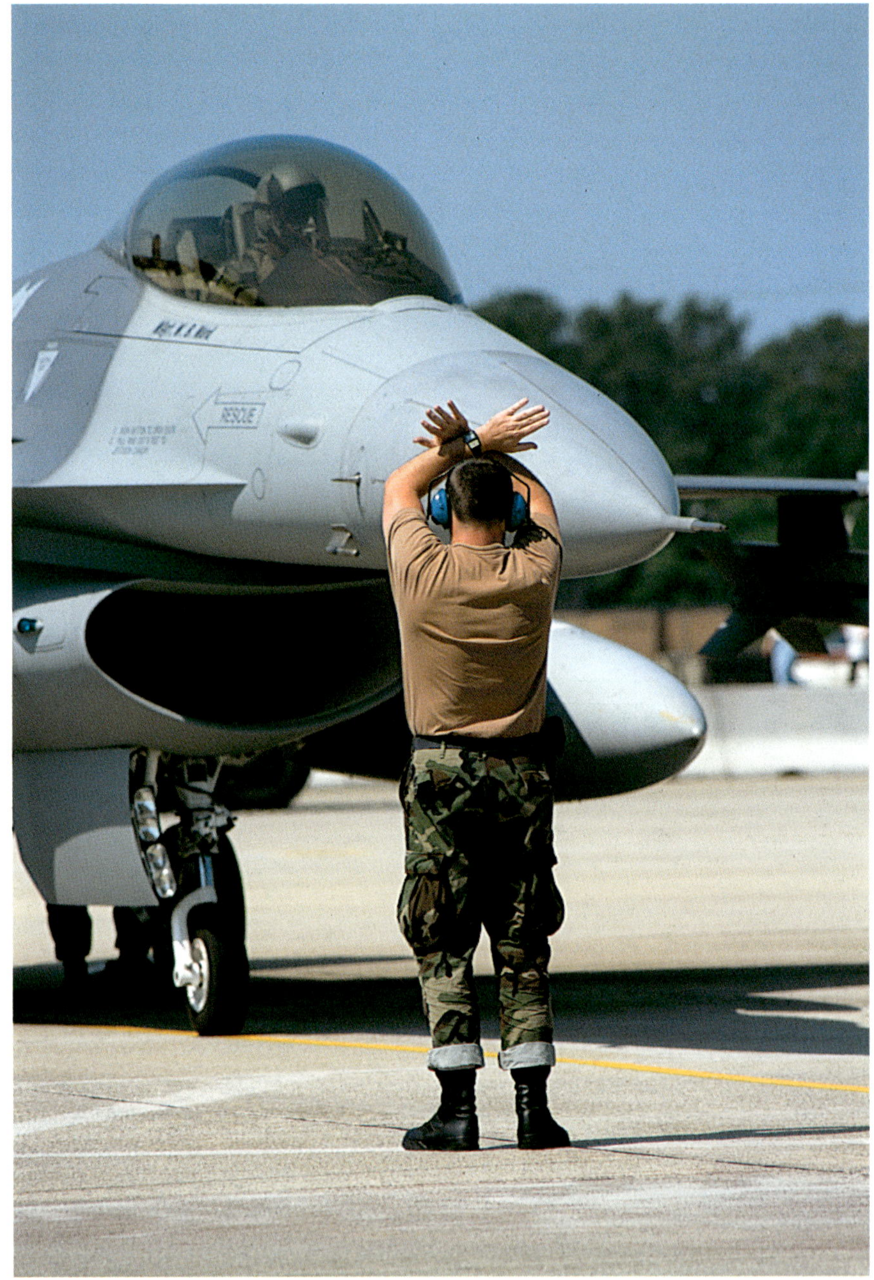

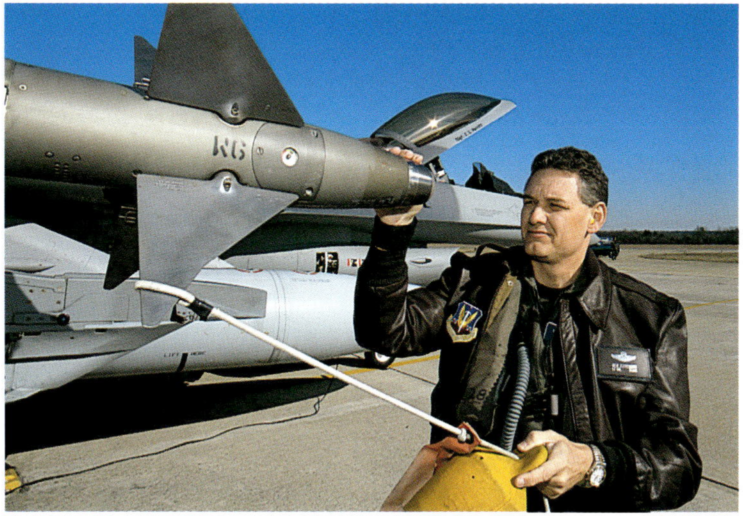

Above
The pilots check the plane and weapons accurately before boarding.

On the left
A 169th Fighter Wing crew chief marshals a Viper to stop on the flight line.

On the right
The success of a mission depends heavily from a good planning and a well-done pre-flight check.

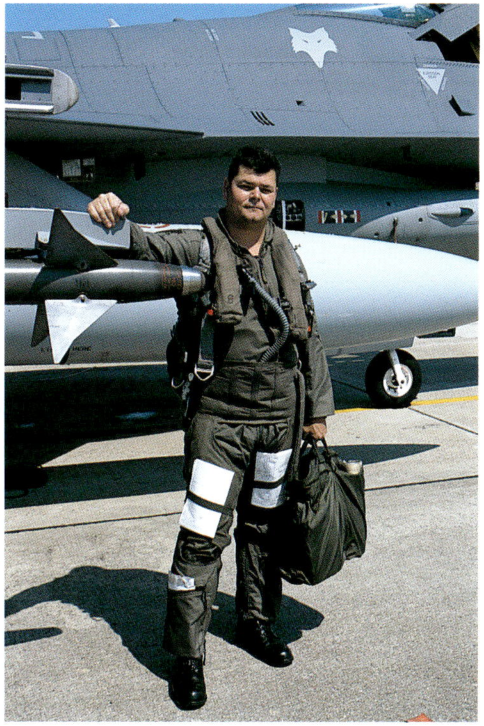

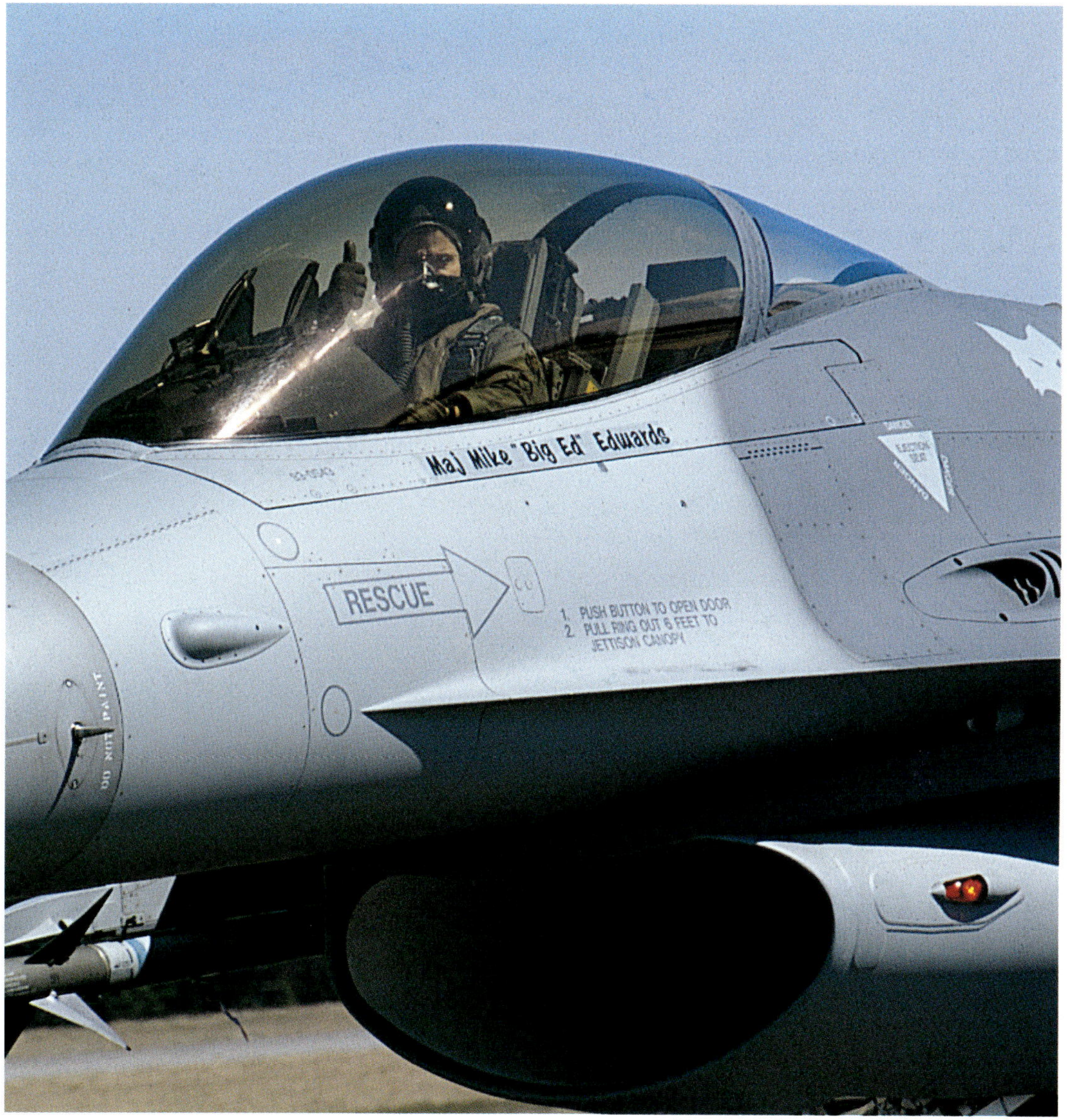

Above
Once the pilot is on board, the thumb-up sign means "all right".

Capt. Thorne Ambrose, an F-16 pilot from Columbia, South Carolina, stressed that "There was a lot of stuff being fired at us - just everyting and the kitchen sink was coming up through the clouds at us...

We were up in target area probably 10 or 15 minutes. It just secmed like times that I was supersonic and it just seemed like I wasn't going anywhere."

Captain A. Wayne Phillips flew in the 157th's second strike package on that day. Near the end of the pre-mission briefing, his target was switched from an airfield in southern Kuwait named Ahmed Al Jabar to several SA-2 SAM sites near that installation. In addition, his take-off time was delayed by a nerve-wracking two hours just as he was climbing into the cockpit of his F-16.

On the right
Reflections on the canopy of an F-16D flown by Lt.Col. Jet Jernigan. Two other F-16Cs fly in formation.

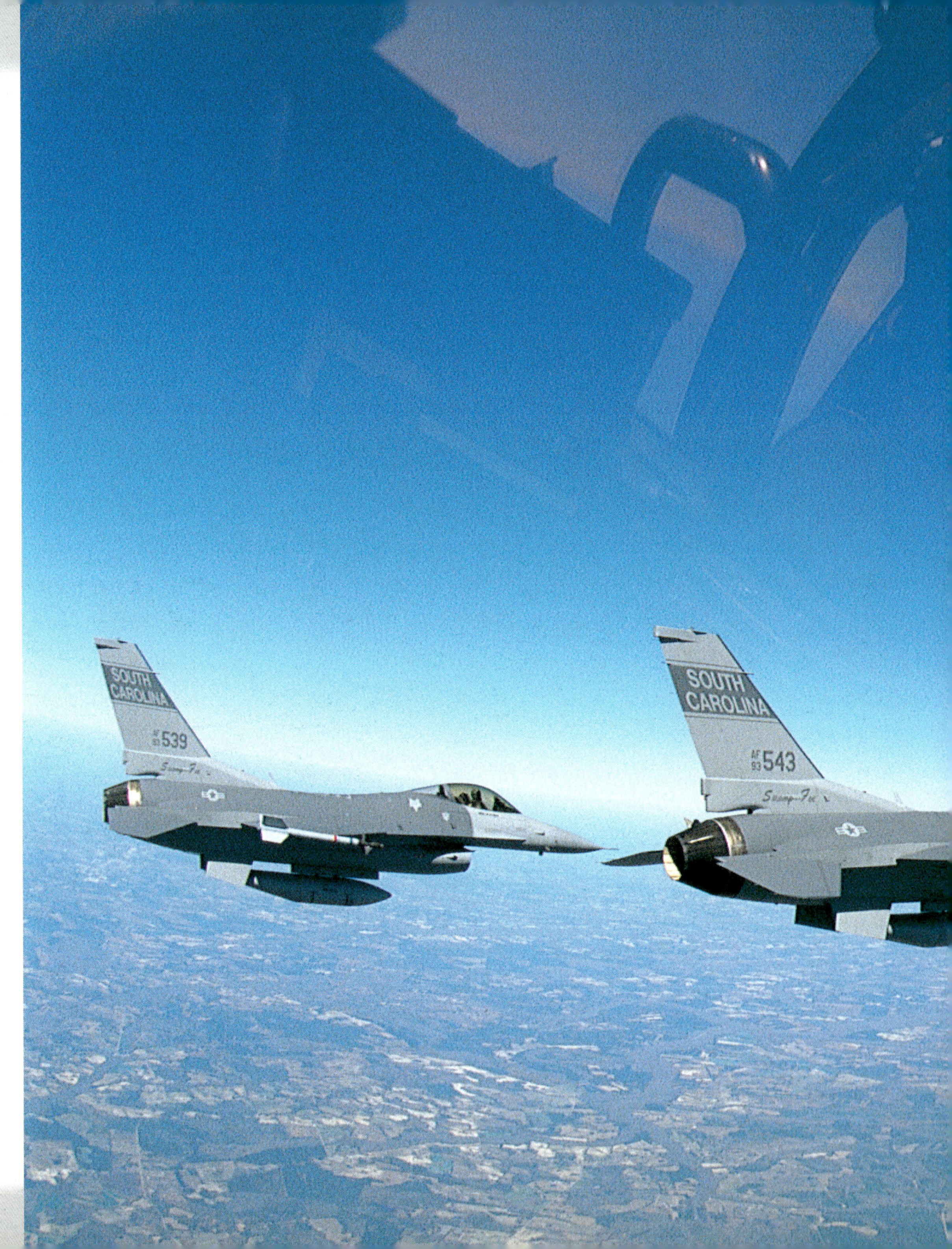

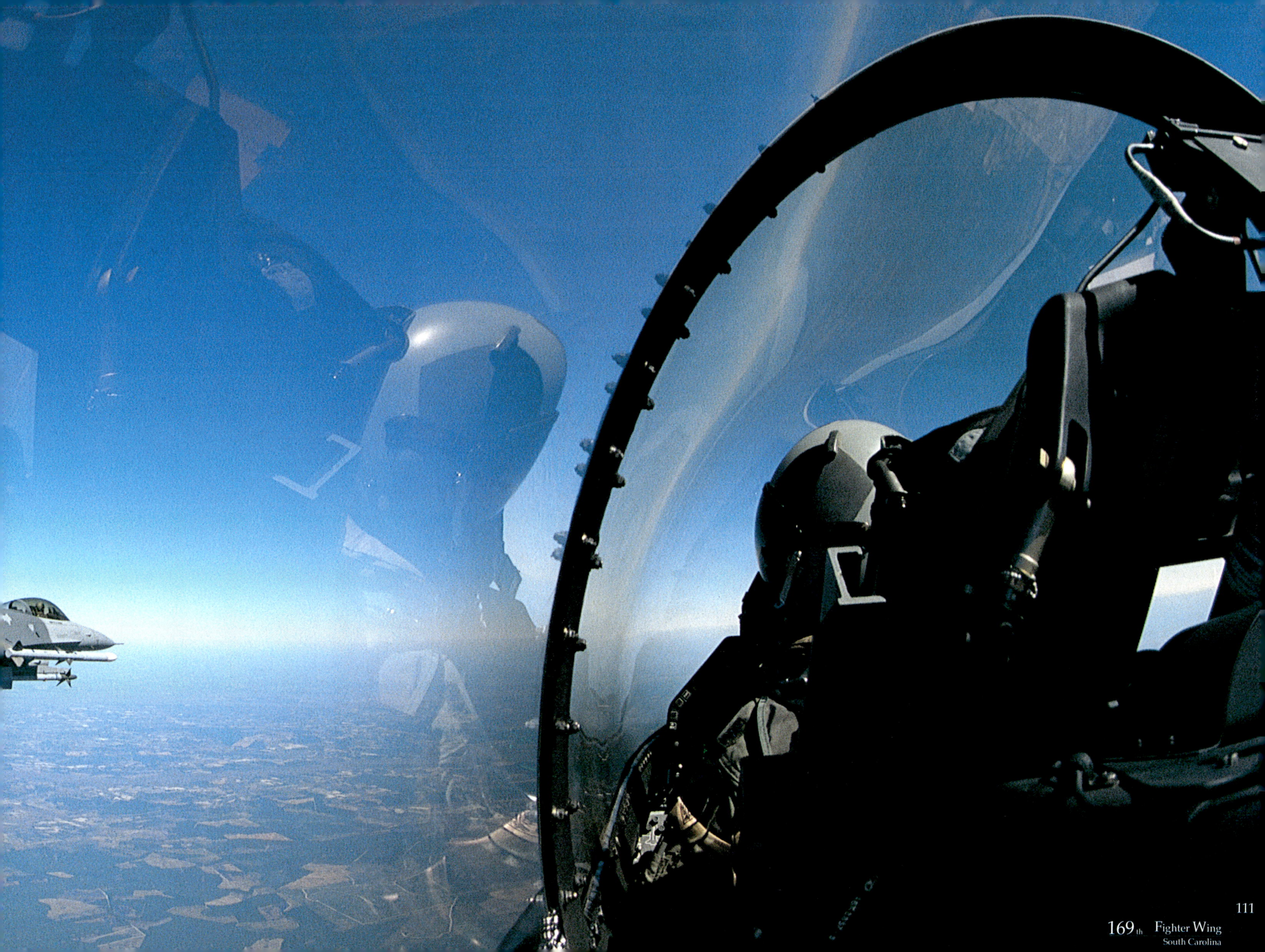

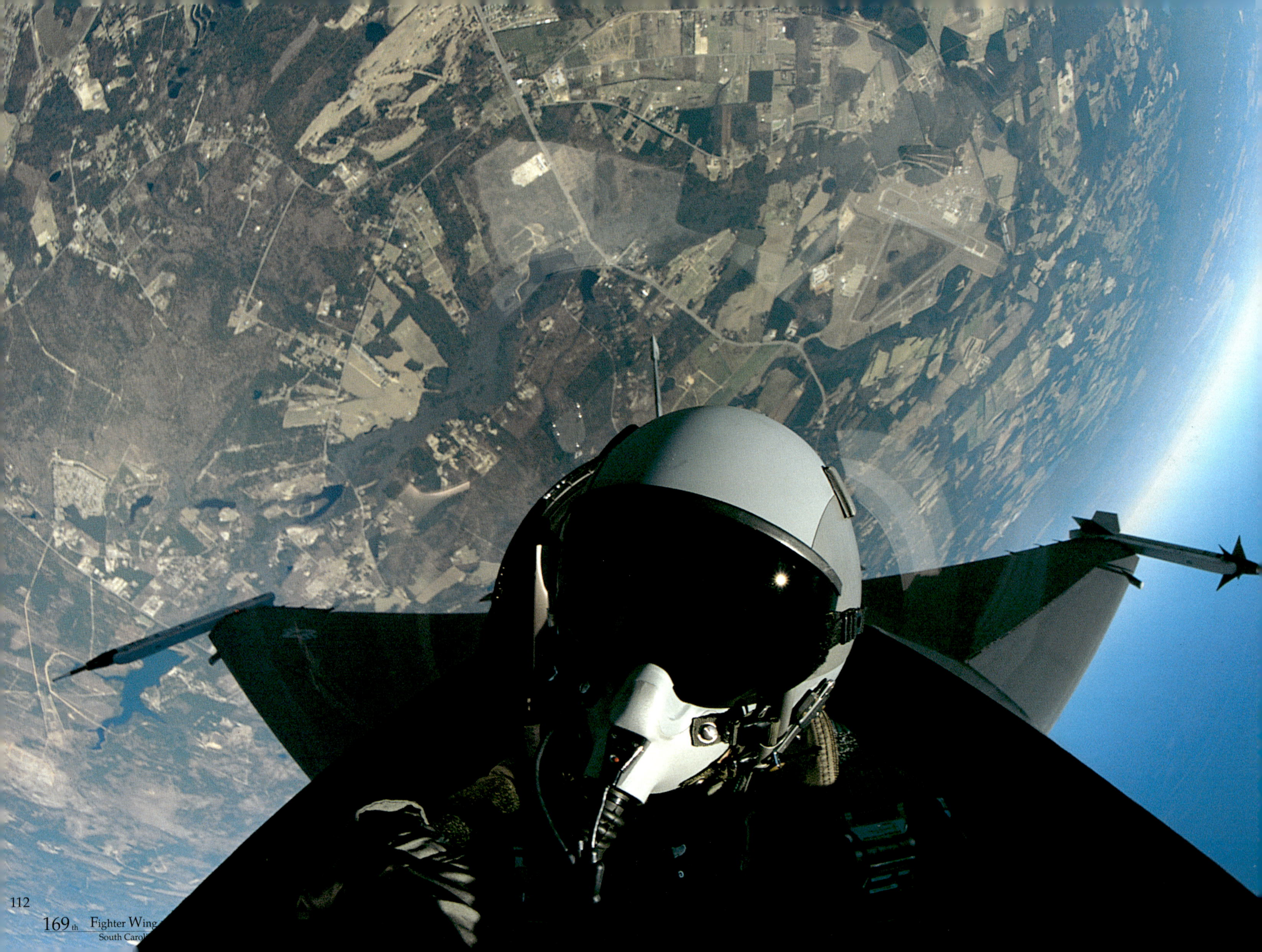

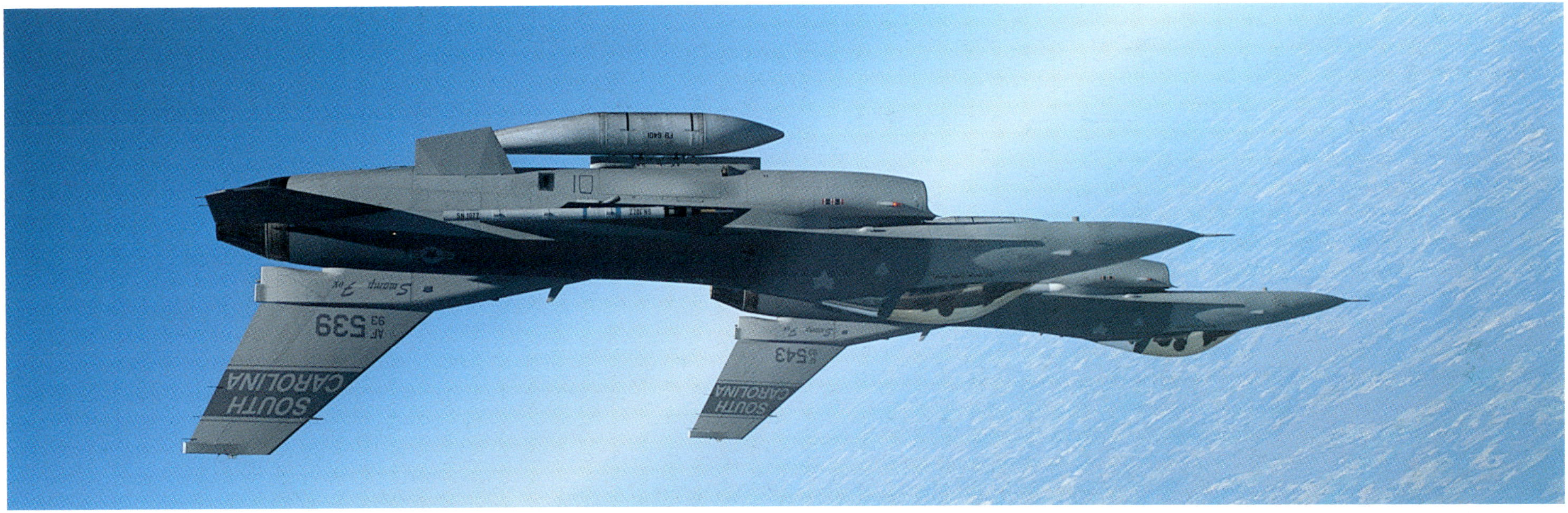

Above
A two-ship close formation "pulls" to close the loop.

In the page on the left
A portrait of a 169th Fighter Wing pilot who makes a chandelle turn atop a steep climb. The air base runways are alwais in his sight.

Although not drinking coffee, Phillips went to bathroom six or eight times. Several times, he had difficulty catching his breath. He and his fellow pilots prepared for the new targets as best they could. Phillips recalled the harrowing details of his first combat mission in his diary: "There were 16 of us and Syracuse 138th TFS, New York ANG also had a 16-ship package... We were the second four-ship of the first flight of 16... We took off right behind Syracuse. When you use a standard 20-second single-ship take-off with live bombs - 8 Mark 84s, 2,000 pounds each - it takes an even five minutes to get 16 ships airborne.

We had F-16s taking off non-stop for over ten minutes. The bad part is that we're getting back at the same time. It'll be dark soon... Weather was a problem. The closer we got to Kuwait, the worse it got. There was a deck at about 10,000 feet, then various layers up to 31,000 feet. We tried it on top but that was a no go - we'd never see the target. We went down to 21,000 feet."

Phillips was uncomfortable with that because it exposed him to greater danger from Iraqui SAMs. His strike package included four F-4G "Wild Weasel" SAM killers and two EF-111 electronic jammers. The Weasels arrived over the target area before the F-16s and reported that the

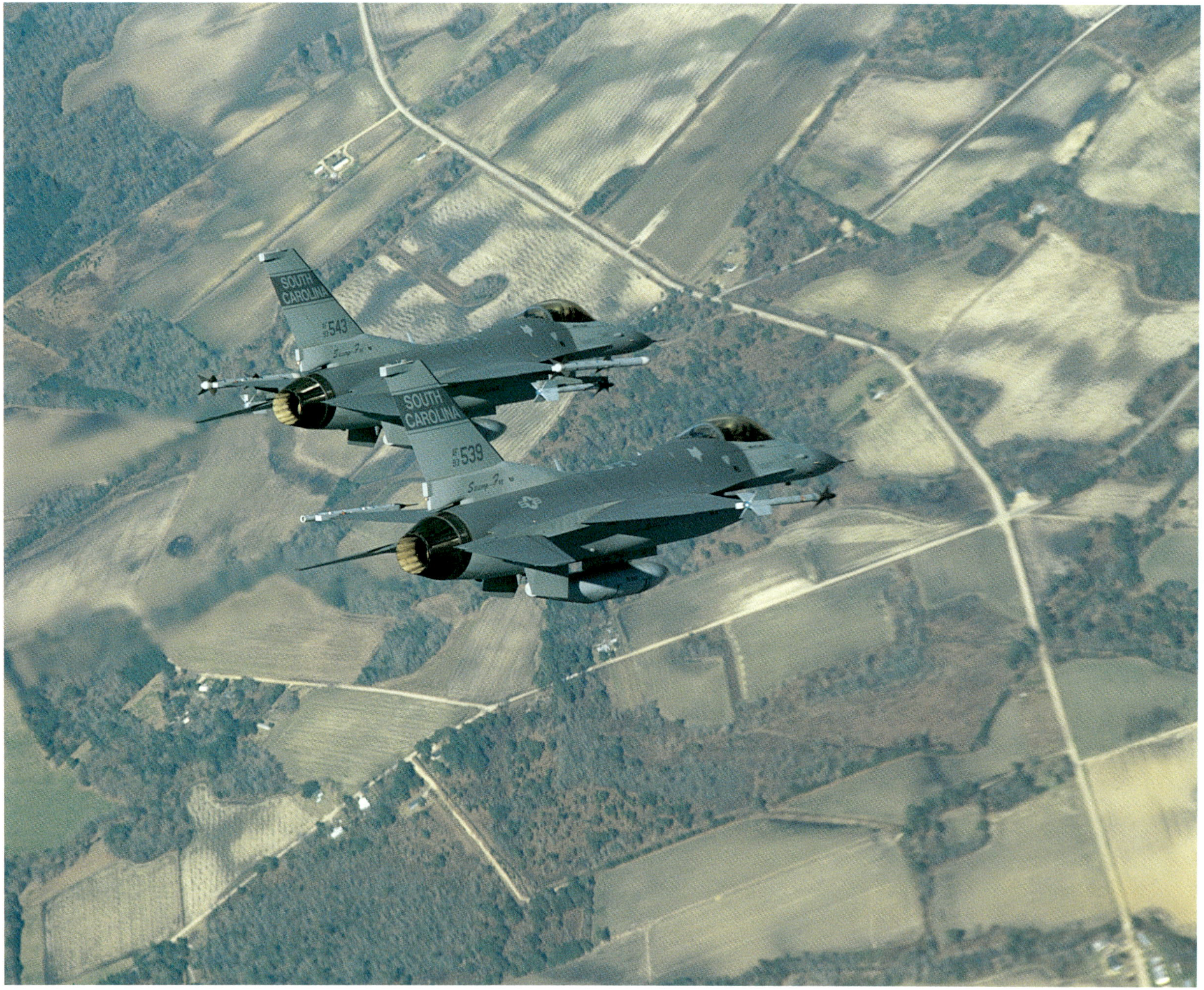

On the left
Two F-16Cs Block 52 dive onto their target.

On the right
Flying in close formation a wingman must check his leader's position and attitude constantly.

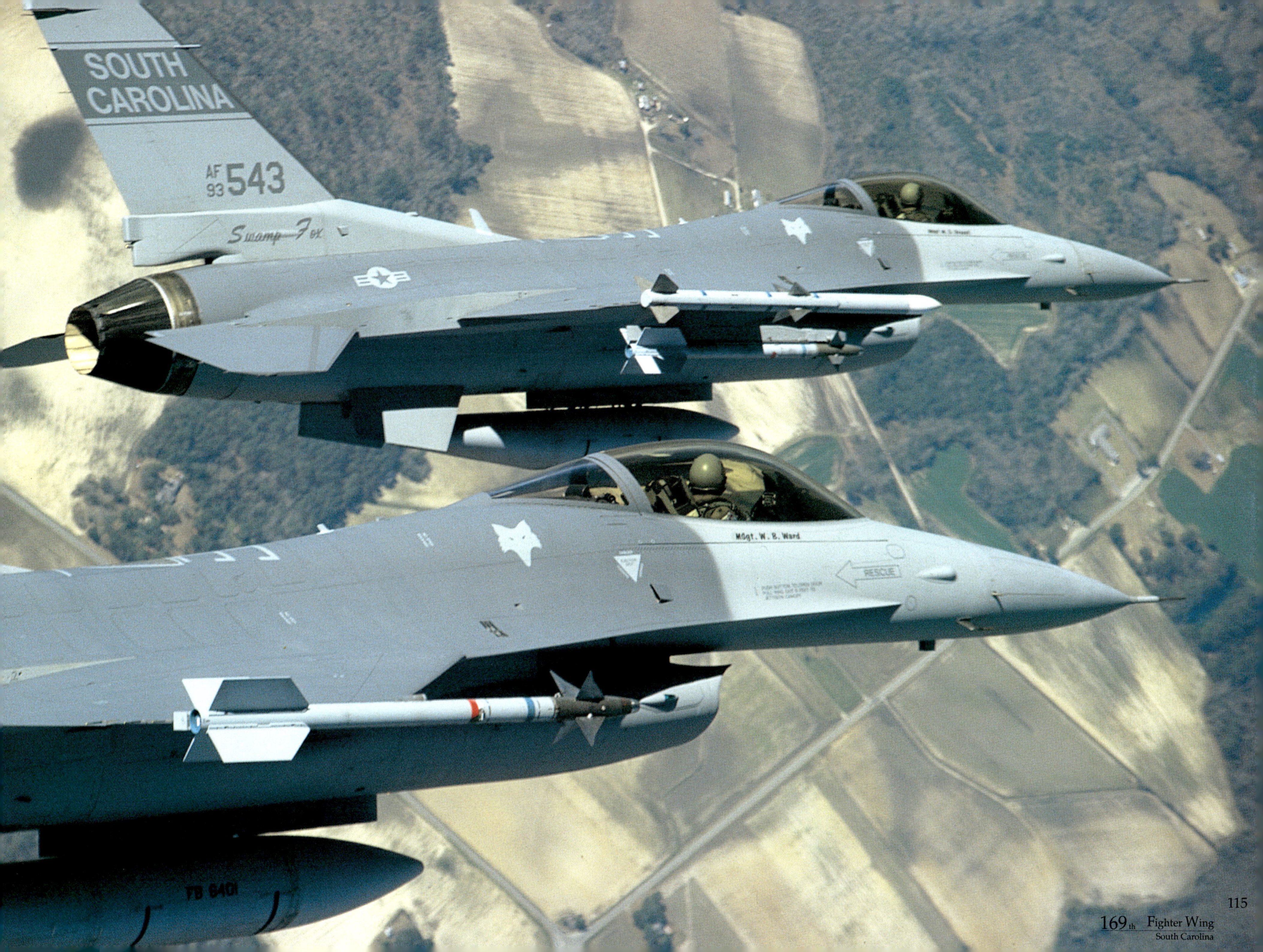
SOUTH
CAROLINA
AF
93 543
Swamp Fox
Msgt. W. B. Ward
RESCUE
TB 8401

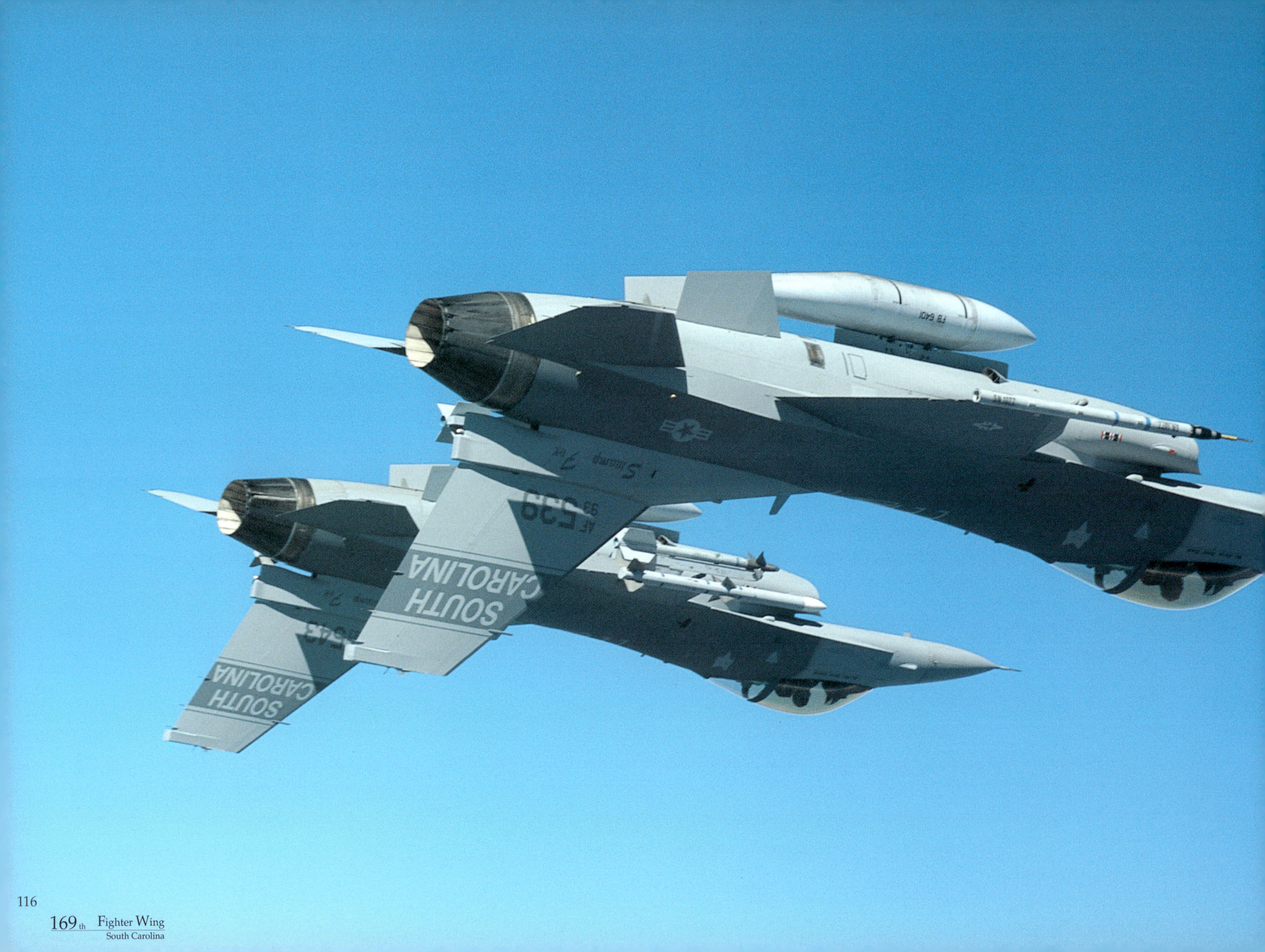
SOUTH CAROLINA
SOUTH CAROLINA

lower cloud deck was thinner and had lots of holes over the target area. Phillips and "Bullet", his flight leader Maj. John N. Bellinger, swung north toward their other two flight members because of clouds and the fact it was twilight.

He recalled that: "Our target was obscured by clouds, so we decided that 'Bullet' would do his best CCRP (Continuously Computed Release Point) delivery, and I would release manually when I saw his bombs come off. With the weather, I couldn't see his bombs, and with 15 other people on the radio, I didn't hear his 'pickle' call (bombs dropped and making a left turn). I pickled and felt two big 'clinks', as my stuff left my wings."

Phillips ignited his afterburner and began a climbing turn toward the South. At that point his radar warnning receiver lit up with missile and anti-aircraft gun indications from all directions. According to Phillips "our egress took us back over the Al Jabar airfield and the eight-ship [flight] was a bit late.

"All of a sudden, there were airplanes everywhere, and SAMs were coming up... [And] the triple A was going off, everything seemed to almost stand still. Something caught my eye... another F-16 crossing in front of my nose!

2At that moment, the South Carolina vowed to concentrate on getting out the target area in one piece. He commenced twisting and popping chaff and got his aircraft going as fast as it could. His main concern was to avoid colliding with another F-16.The return flight was a nightmare."

Phillips stressed that "As we went in and out of the weather and things got dark, I lost sight of the others. I went to 35,000 feet and put Saudi Arabia on my nose and ran like a bat out of hell until I was sure I was back in 'good guy land'."

Although the rest of the trip was relatively uneventful, Phillips was still tense. He had to grope through the night to Al Kharj, a field where he had never landed before after dark, with 30 other Air Guard F-16s. Somehow, they all made it back safely.

After shutting down his engine and opening his canopy, Phillips got a big hug and an Iraqi flag from his crew chief. Following a debriefing, he headed for the chow hall. His reward for his first combat sortie was a tuna casserole.

In the page on the left
The formation aerobatics are a basic drill for air-to-air combat tactics training.

In the page overleaf
A two-ship formation of South Carolina ANG Vipers banks on a turn.

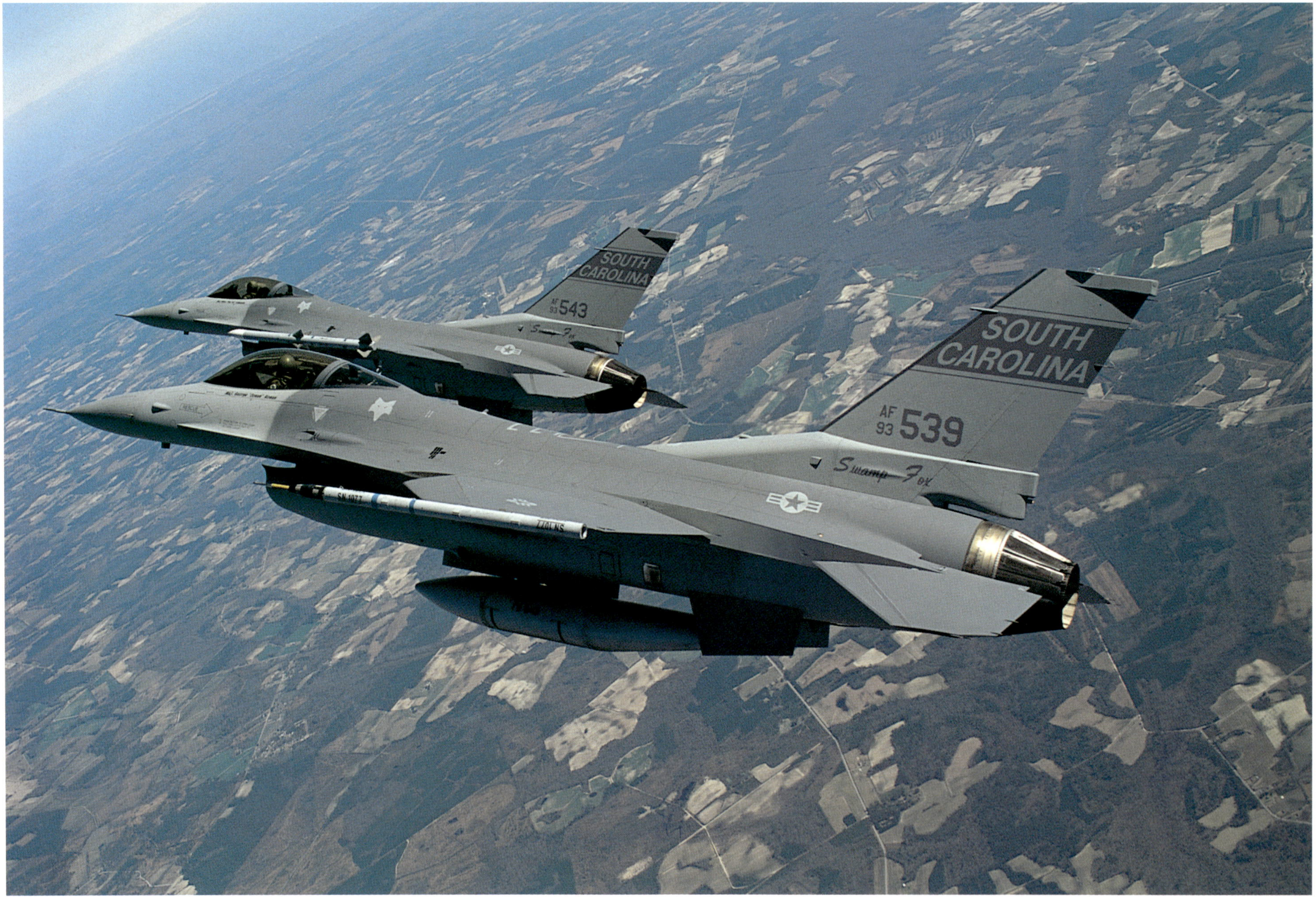
SOUTH CAROLINA
AF 93 543
SOUTH CAROLINA
AF 93 539
Swamp Fox

180th Fighter Wing

Ohio

180th Fighter Wing, Ohio (emblem approved 20 July 1965). The equilateral triangle, 180 degrees, represents the unit's numerical designation and the stars, the subordinate units. The flight symbols, lightning flashes, and globe represent the unit's flying mission, teamwork, coordination, speed and power in world-wide operations. The colors represent the Air Force yellow and blue and the national colors. The unit motto is "Nos Cives Ultro Milites" (We Citizens Serve Voluntarily).

112th Fighter Squadron, Ohio (emblem approved 9 March 1954). The bee is social in its habits and lives in a community where each member participates in some service for the common welfare of all, as the Guardsman lives in his community and participates in ensuring the security of the community and the nation. The sting of the bee, like the fighter squadron, is primarily a defensive measure. The armament carried by the bee represents the firepower of the aircraft, which is always ready to defend community and nation from aggression.

112th Fighter Squadron, Ohio. Organized at Kelly Field, Texas on 18 August 1917 as the 112th Aero Squadron, Air Service, this supply squadron was redesignated as the 633d Aero Squadron on 1 February 1918. It remained at Kelly Field and was inactivated on 19 August 1919.

In 1936 its lineages and honors were consolidated with the 112th Observation Squadron, activated as an Ohio National Guard unit on 20 June 1927 at Hopkins Airport, Cleveland. It served as an observation unit for the 37th Division, Ohio National Guard from 1927 to 1940. Mobilized on 25 November 1940, it remained in the continental U.S. until shipped to England in May 1944. The unit flew anti-submarine patrols along the Atlantic coast until reorganized as the 112th Liaison Squadron.

Equipped with Cessna UC-78s it flew courier missions in Europe between June 1944 and VE Day. The unit was redesignated as the 112th Bomber Squadron and allotted to the Ohio National Guard on 24 May 1946.

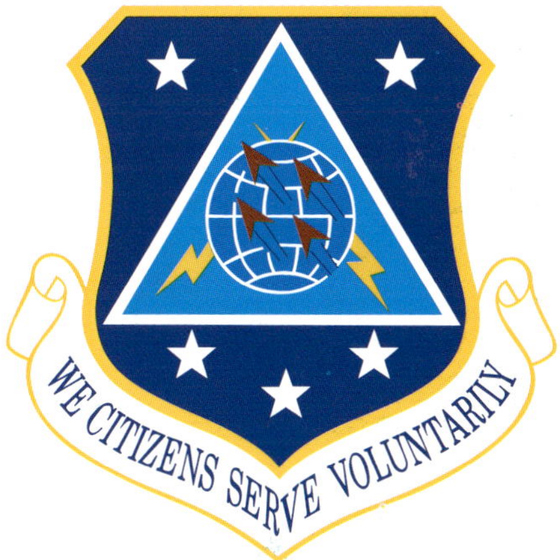

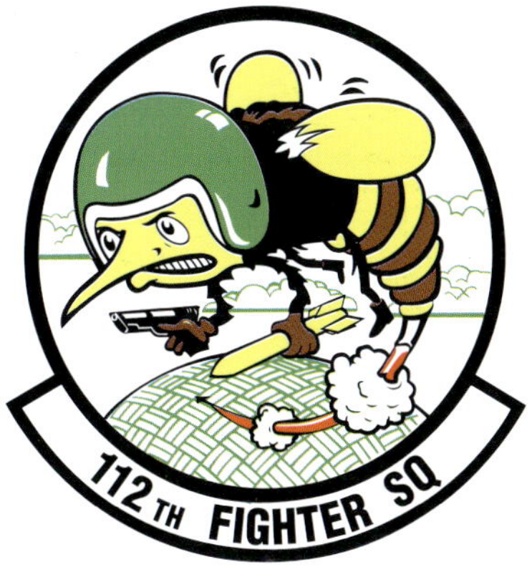

Equipped with Douglas A-26B/C Invaders, the unit was mobilized for the Korean War on 10 October 1950 and moved to Lawson AFB, Georgia. Returned to state control 9 July 1952, the unit was equipped with F-51Hs, it then moved to Akron-Canton municipal airport in North Canton, Ohio.

On 1 October 1961 the squadron was mobilized for the Berlin Crisis but remained on station at Toledo Express Airport, Ohio. It returned to state control the on 20 August 1962 and gained the 180th Tactical Fighter Group as its headquarters on 15 October of the same year. The unit converted to many different kinds of aircraft to include: Lockheed T 33A; Republic F-84E and F-84F; North American F-100D; and then to the Vought A-7D in 1979. In December 1989 the unit flew 22 combat sorties during the invasion of Panama, Operation *Just Cause*. It converted finally to F-16 from 1992-1993.

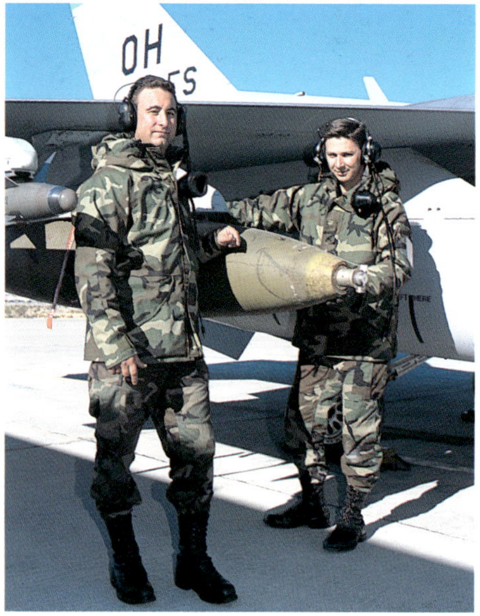

On the left
Two weapons specialists check the fuze of a general-purpose Mk 82 bomb.

Below
An F-16C Block 42 on the flight line is readied for a training mission with "live" ammunition.

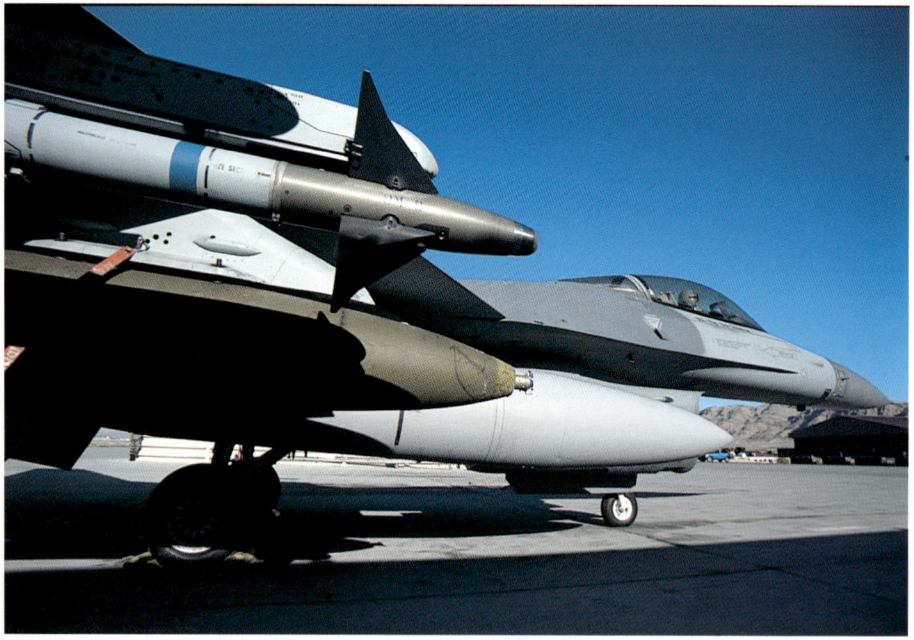

THAT DAY AT NELLIS

BY GIAMPAOLO AGOSTINELLI

Nellis, near Las Vegas, Nevada, is an air base with a history and a lure unique in the United States Air Force. It is often used, and it is appreciated, because of the possibility to train with real weapons, being the base close to several firing and bomb ranges and reserved airspaces for very low-level flight over deserted areas.

At Nellis is the Weapons School, where weapons instructors and tactical operations instructors (including adversarys' tactics) are formed and qualified. Another goal is added to this primary task: to contribute to the upgrading of the combat capability of the United States Air Force, the Air National Guard and the Air Force Reserve by means of various courses, clinics, counseling and specific training programs.

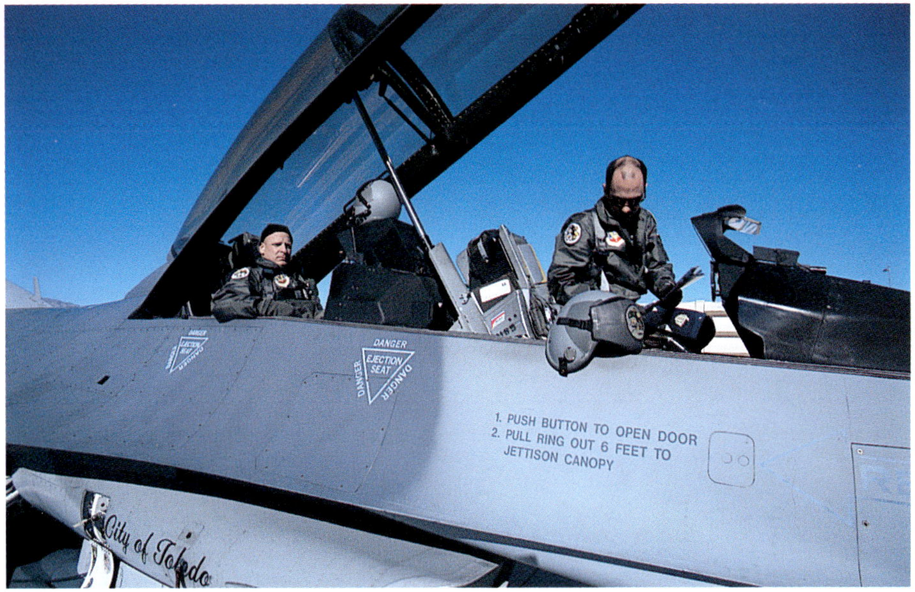

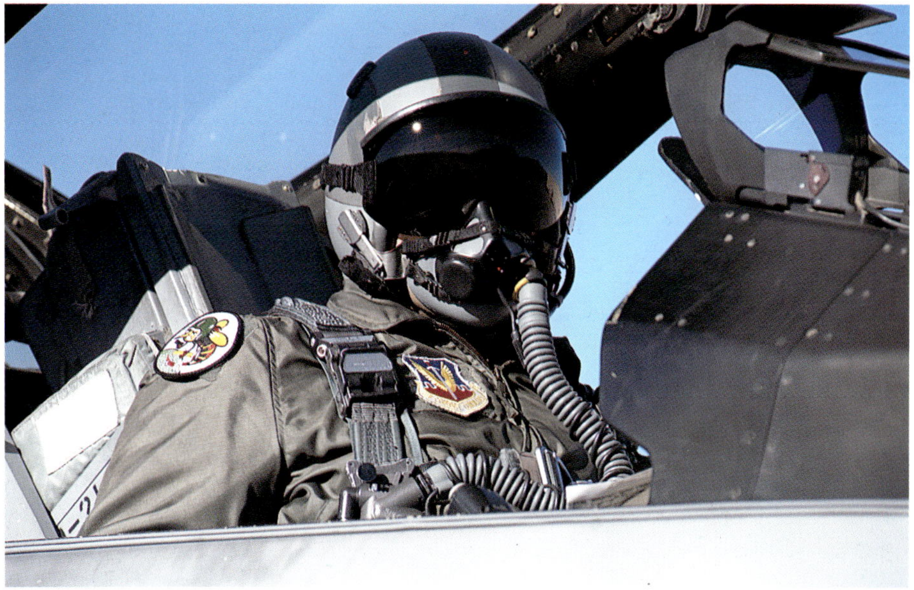

On top, left
Lt. Kelly Connel and Capt. Walt Erold take place in an F-16D for a sortie from Nellis AFB, Nevada.

And, on the right
Capt. Chris Belli aboard his F-16C.

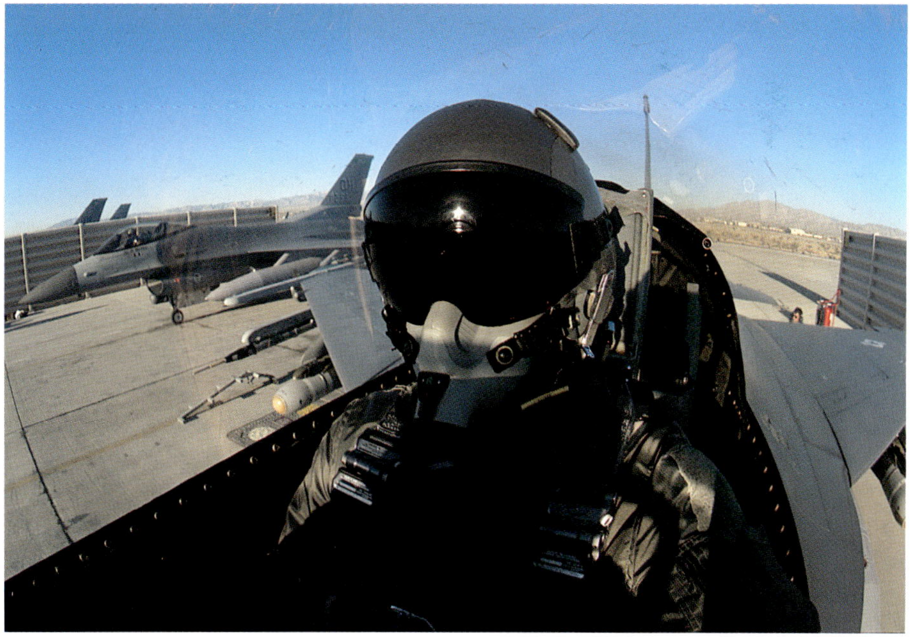

On the right
The author sitting in the reas seat of an F-16D armed with two Maverick air-to-surface missiles.

In the next page, oveleaf
Two F-16Cs diving over the bombing range, north of Death Valley, California. Each one of them carries two GBU-10s

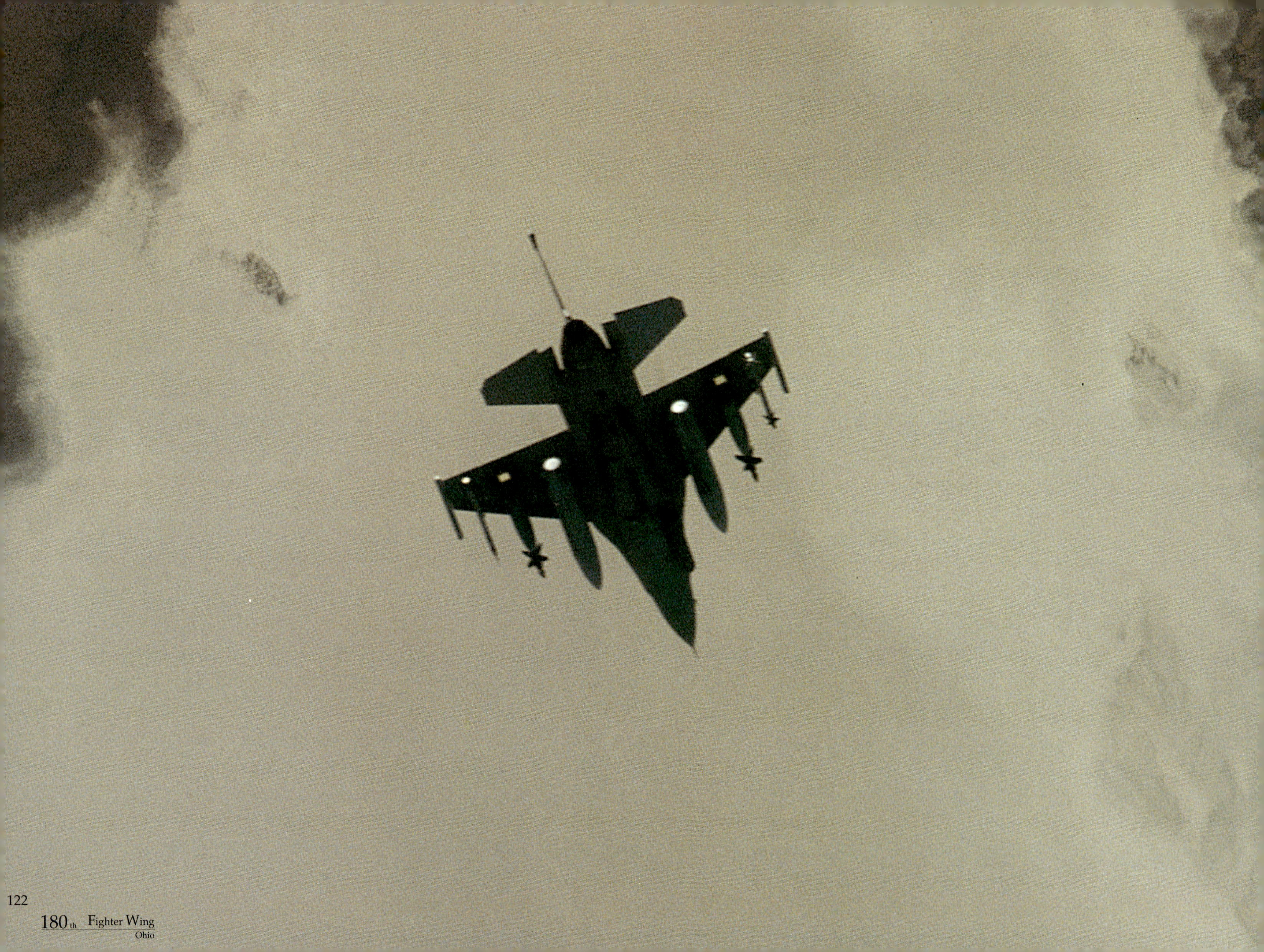

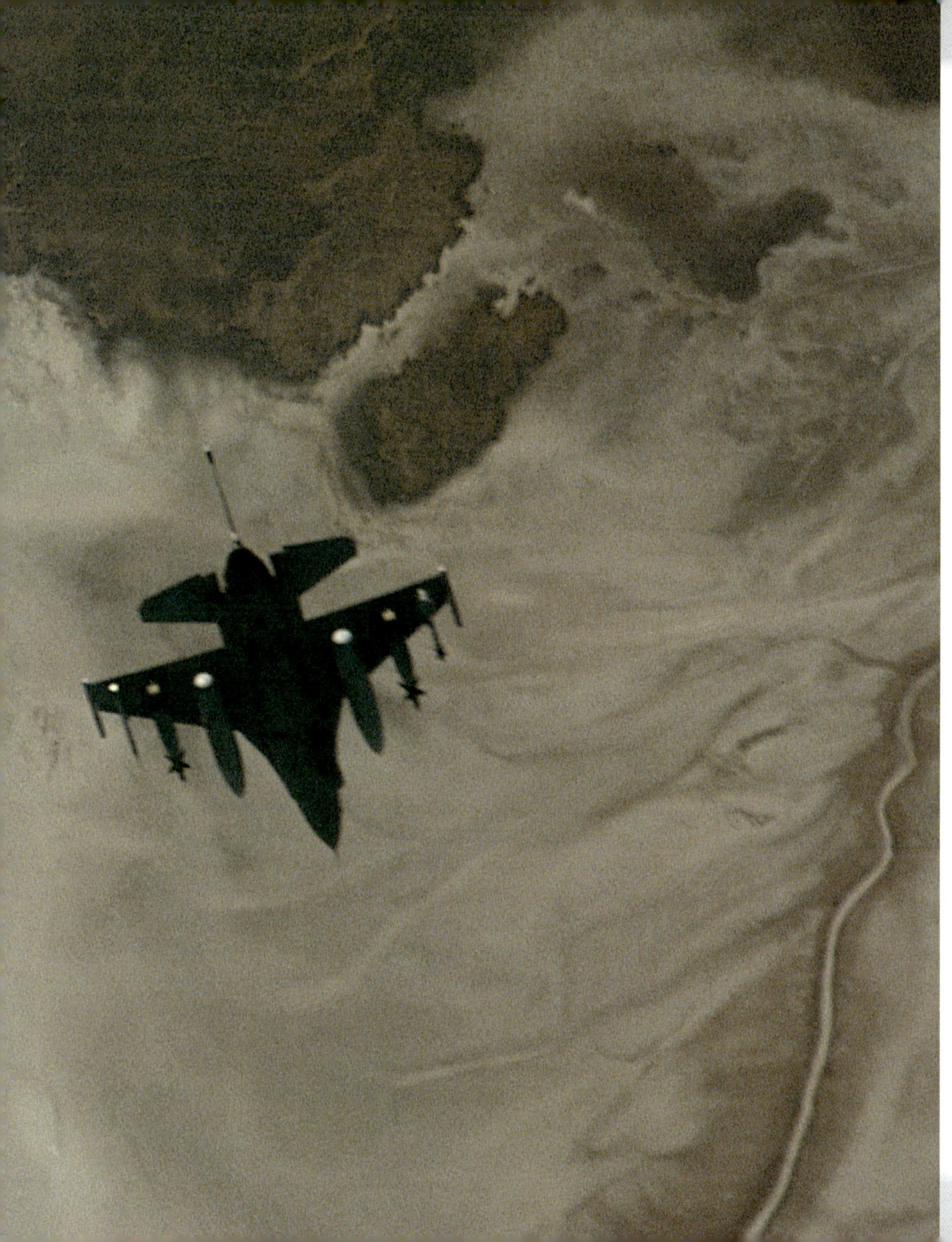

Inside this structure at Nellis is the 549th Combat Training Squadron "Air Warrior", a special training unit whose main scope is upgrading, prepare and update the crews for tactical support aircraft; the whole in a particularly realistic scenery.

The aircraft operate from Nellis, but perform their mission in the US Army National Training Center, Fort Irwin, California, close to the Nevada border. It is early morning, when we meet in a briefing room in the "Air Warrior" building. It's 06:15 am. I find 180th Fighter Wing pilots, Capt. Joe Maslar, in charge of leading the mission, Lt. Kelly Connel and Capt. Chris Belli, who will fly the same F-16D in which I flew at Toledo. The mission will be a real bombing by means of real GBU-10 laser-guided bombs.

This 2,000-lb weapon is one of the four variants of the Paveway II, the others being the 500-lb GBU-12, the 1,000-lb GBU-16 and the 1,000-lb Mk. 13/18. In production since 1976, the Paveway II was refurbished and upgraded by Texas Instruments in 1990. The GBU-10 has a sinking speed averaging 1,000 feet per second and it is possible to correct its trajectory some ten times in its final flight to the target.

The war scenery of Operation Desert Storm, which represents the type of conflict selected as a reference, highlighted the growing importance of precision strike. The pilot who is lucky enough to complete a training cycle at

"Air Warrior" gets an unequalled knowledge, refining his flying tactics and becoming aware of the full potential and limits of his aircraft. This is of vital importance in view of having units always ready for deployment in "hot spots", and able to operate immediately and effectively.

The 112th Fighter Squadron will spend some three weeks at Nellis for an "Air Warrior" course to improve in the use of Maverick air-to-ground missiles and of free-fall and guided bombs, in preparation to the Northern Watch exercise that will be held between September and October.

"We set the time in our watches: three... two...one. It's 06:30 now" says Capt. Maslar. The mission plan and all emergency procedures are discussed in detail, in case something does not run in the right sense.

We stop at the flying gear shop: at 07:10 we are ready to step to the planes. A van carries us to the South Base, to a smaller but well equipped apron where real weapons can be handled in a safer way. Chris performs the external routine check while I ease myself in the rear cockpit in order to take enough time for having all my gear prepared. Engines are started up at 07:40. The hissing sound of the Pratt & Whitney F100 is barely heard in the soften atmosphere under the wide transparent canopy.

"Pre-taxi checklist" announces my pilot. "With real weapons we must check everything much more carefully". Beside the aircraft crew chiefs I see the weapons specialists, busy with their cross checks. Meanwhile Chris invites me to check my flight controls to full deflection. Everything appears working properly.

"OK, we go: also the targeting pod is ready". The most important device that we are going to use in our mission is the laser targeting system which guides the "smart" bombs onto their target. The targeting pod of the F-16 is the Lockheed Martin AN/AAQ-14. Integrated in the weapons control system and with the INS, it allows various modes of target acquisition and attack. Slung under the right side of the air intake, this pod weighs 245 kgs and is 250 ins long and 38 wide. It contains a laser emitter (the same which guides the bombs onto the target), a FLIR, a data processor, an electronic control unit, a power generator and an environment control unit. The system can be used both for air-to-ground and air-to-air attacks.

The search head, which contains the FLIR and the laser emitter and receiver, is stabilized in a way to compensate for the aircraft's movements and vibrations. In the central body, just behind the mobile "nose", all the electronics for target determination and data link to bombs and missiles find their room. After having stopped at a taxiway turnout where our weapons safety pins are removed by the specialists, we are ready for take-off.

In the page on the right
A Viper starts its run-in, acelerating at low level before "popping-up" to altitude to release its two precision-guided bombs.

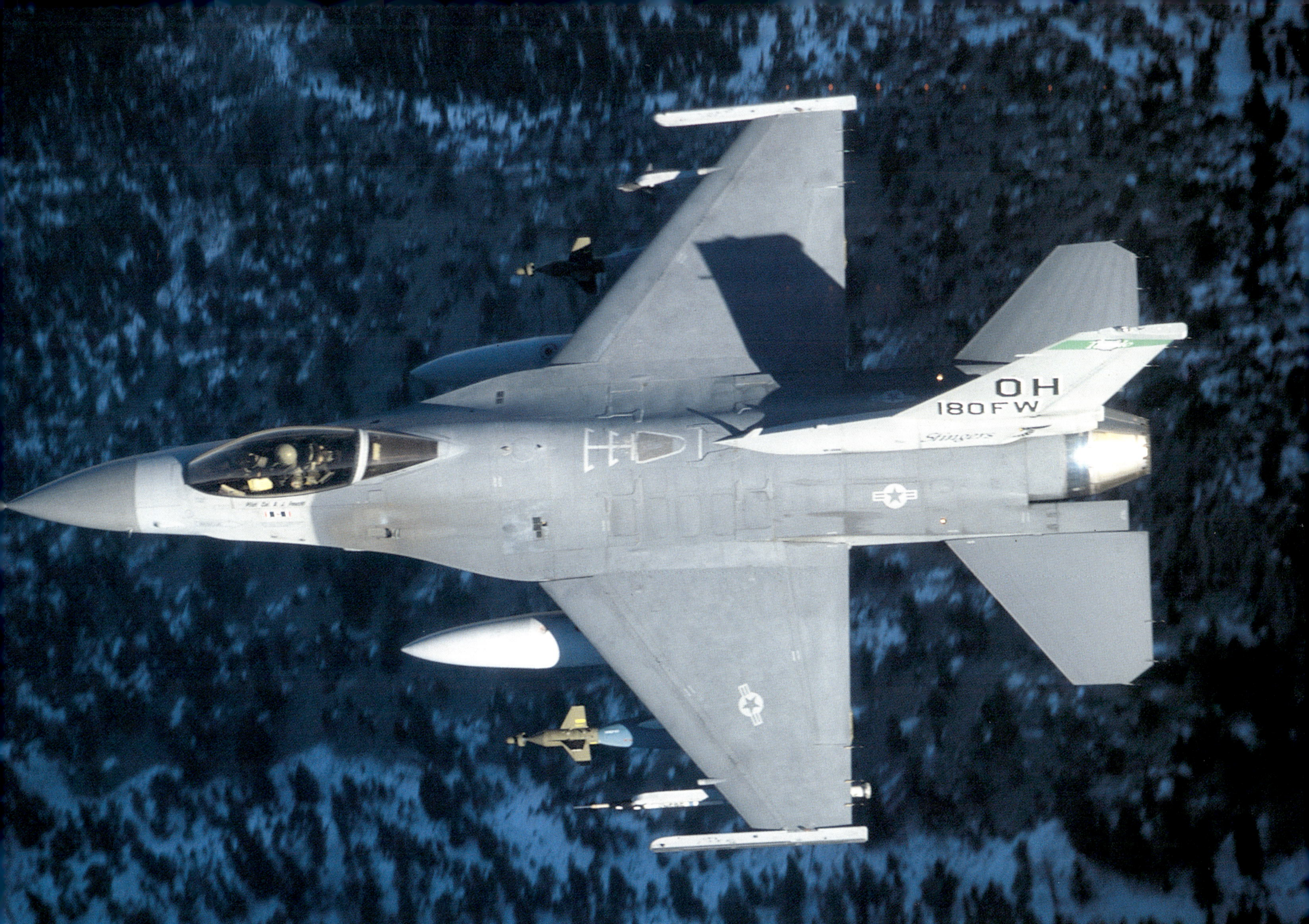
OH
180FW

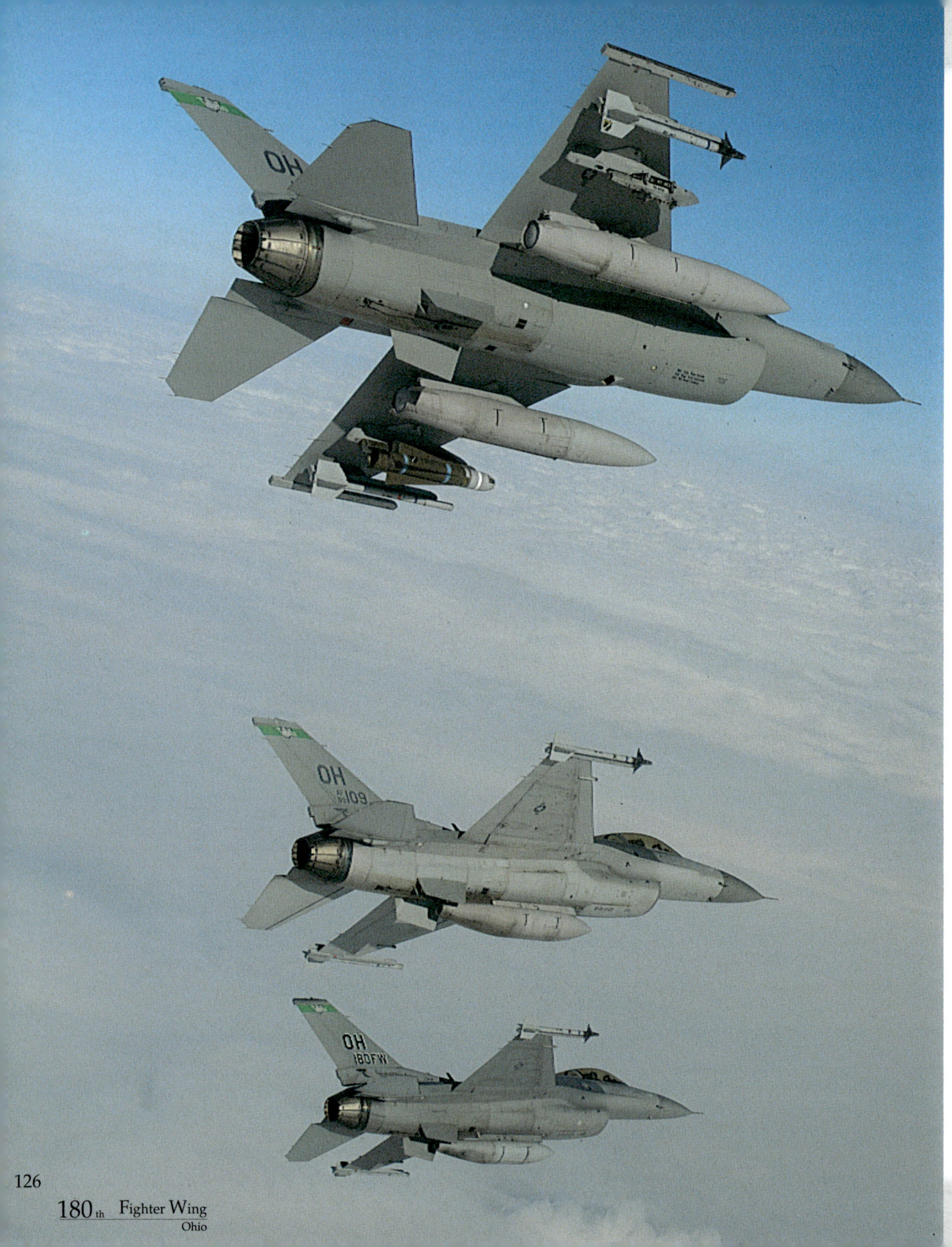

With 8 seconds separation, the two single-seaters roll first, then it's our turn. With "full AB" on, we lift off and immediately start an 180 degrees turn to join the tactical formation. We proceed at low altitude above the ground and enter a narrow valley through a ridge north-west of Las Vegas. There we enter the "foe's land".

Our targets are an AAA battery and an adjacent bunker, south-west of a simulated Iraqi air base. As the bombing range is so close to our departure base, we are going to simulate a longer flight to the target, passing several waypoints over deserts and ridges. In Nevada are spots of land where mission profiles against Iraqi targets can be very well simulated.

The low-level flight in which the three F-16s maneuver in a way to cover-up from the adversary's radar sweep, allows us to get to the back of our target, where we use a nearby ridge as a final screen.

On the left
Echelon turn or three F-16Cs Block 42 of 112nd Fighter Squadron on their way back from a range in the Nevada desert.

In the page on the right
Neat and compact lines and mighty appearance - this is the Viper.

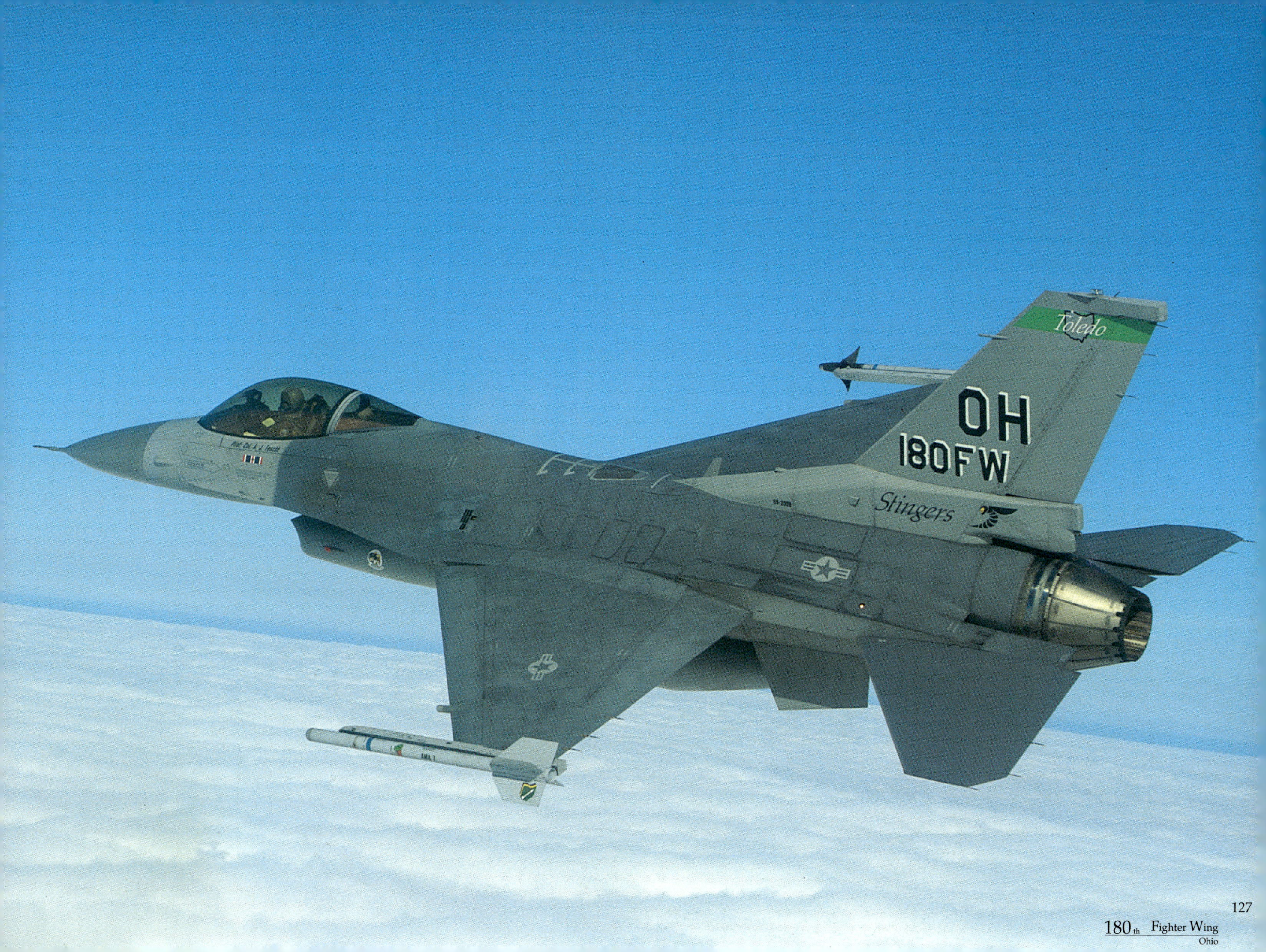
Toledo
OH
180FW
Stingers

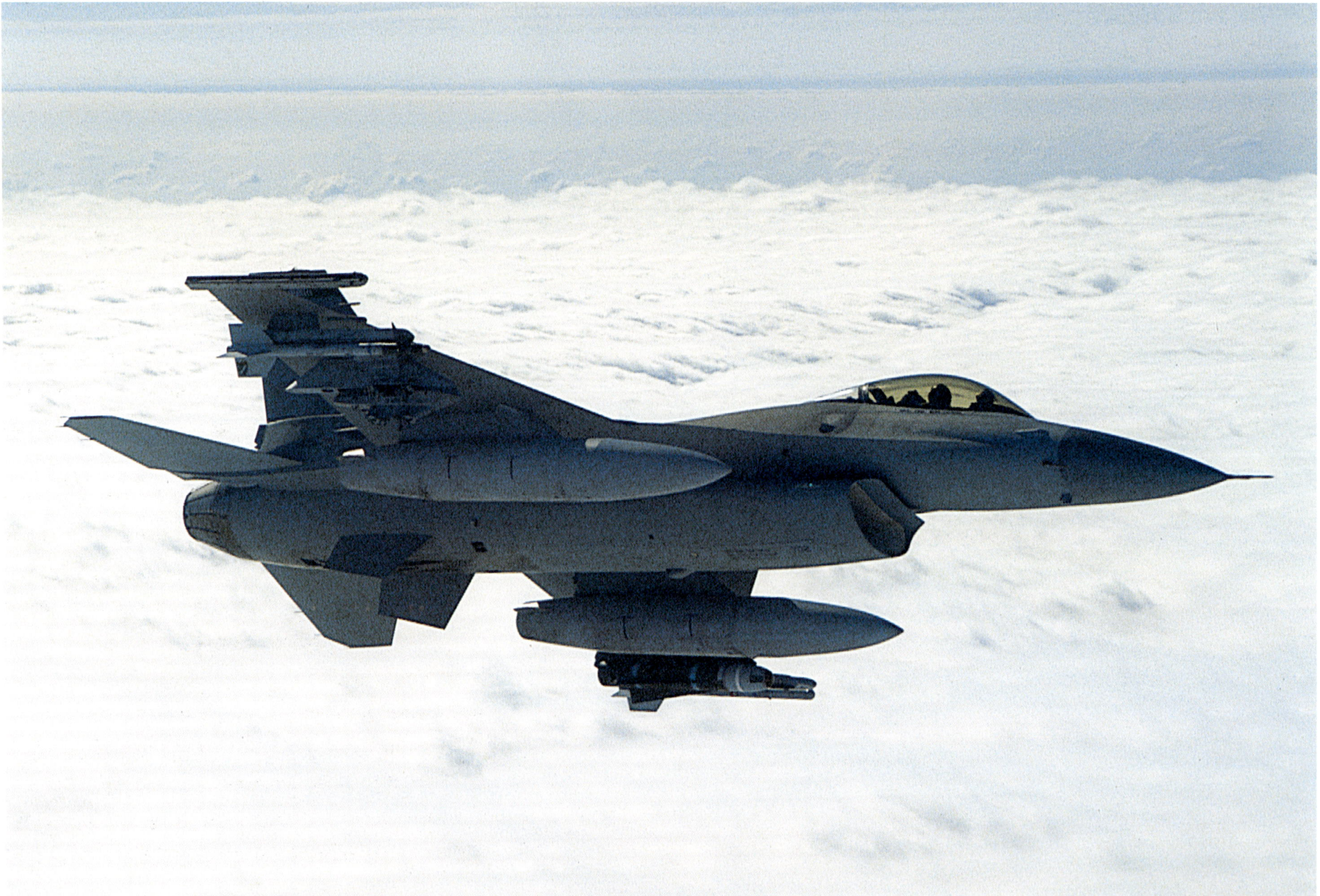

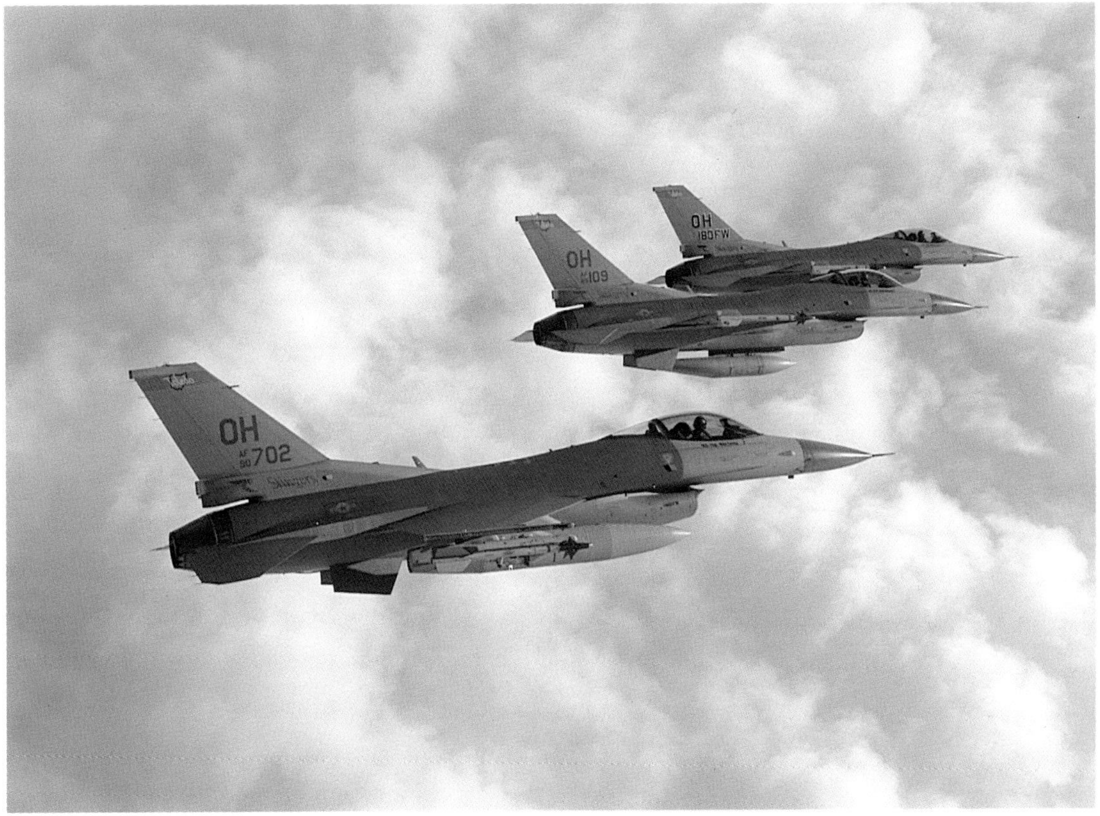

Above
An unfavourable forecast does not dissuade 112nd Fighter Squadron pilots from their usual week-end drill.

In the page on the left
A Viper rejoins the formation on top after a last pass over the firing range.

Everything had been studied accurateley on the eve's evening and rehearsed in the briefing. Thanks to the detailed maps and updated satellite photos of the target at our disposal, we are proceeding with safe determination in our mission. The target coordinates and the run-in and run-out waypoints have been punched in the DTU, Data Transfer Unit in the Ops Room.
Nothing is left to improvisation: at each waypoint, position and time have been checked precisely. Now we start the run-in behind the last ridge. Suddenly we climb to 11,000 ft: our targeting pod is "hot". I can follow the target acquisition phase on the right-hand Head-Down Display, set in high-resolution radar mode. I can actually "see" the bunker, our target.
With a little roller on his throttle handle, the pilot drives the cursor on his radar display until it matches with the target return. The closely-spaced targets assigned to our mission add some complication to his job, as he has a very short time to lock-on his own and to verify the target pod response.
All has been planned in order to produce the maximum difficult situation and to allow us to learn to manage it and solve the related problems.
OK: the cursor is on the right target and locked on it. The targeting pod responds positively at 23-mile range, 303

KIAS (knots, indicated air speed), 10,800 feet. I can see the entire situation on my Head-Down Display.
Now everything runs automatically: the pod has started the release countdown for the two GBU-10s - 15 seconds from now, as set - keeping the aircraft straight and level to the target.
Five-four-three-two-one. Go! They have gone!
Our targeting pod made a good job, and we start a quick and "heavy" right turn, beginning our evasive maneuvers to leave the target area as fast as we can, avoiding the adversary's defensive reaction.
Meanwhile, the AN/AAQ-14 head remains locked on the target, illuminating it for the search heads of the GBU-10s released a few seconds ago. My right-hand HDD continues to give me a picture of the target area, and I can see the two flashing objects hit the bunker and explode in a big fireball and dense black smoke.
"OK, that's bull's eye" Chris confirms. On our return trip, we relax for a while commenting the well done job.
That's Nellis on the horizon. We slow down for initial approach, then we break to the left to enter the traffic pattern. One minute to touch down.
The runway comes in very fast in the hot air. Today I have added an important and substantial part to my flight experience.

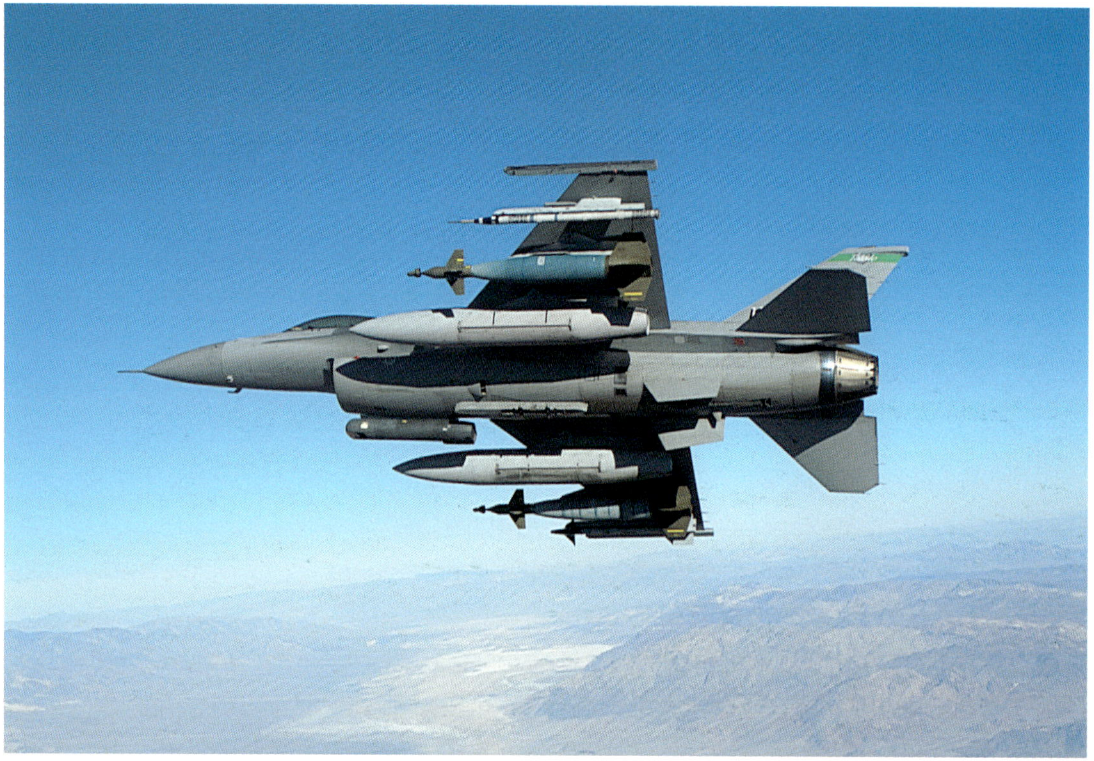

Above
An F-16C Block 42 armed with two GBU-10 (practice) precision bombs and equipped with a Lockheed Martin AN/AAQ-14 targeting pod turns onto the bombing range..

In the page on the right
Two laser-guided GBU-10 practice bombs are released by an F 16C.

In the next page, overleaf
A Viper prepares for descent in instrument flying conditions.

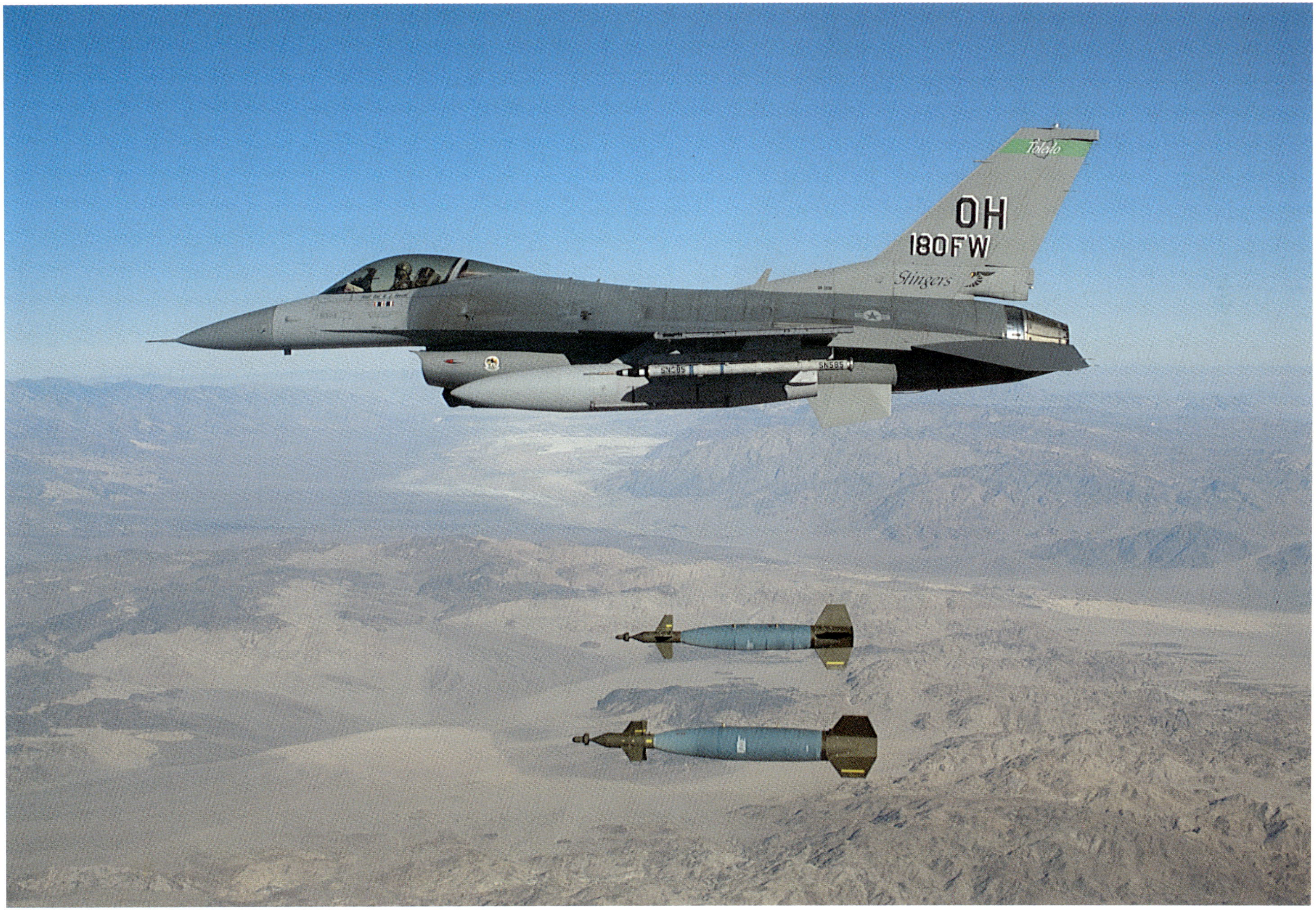
Toledo
OH
180FW
Stingers

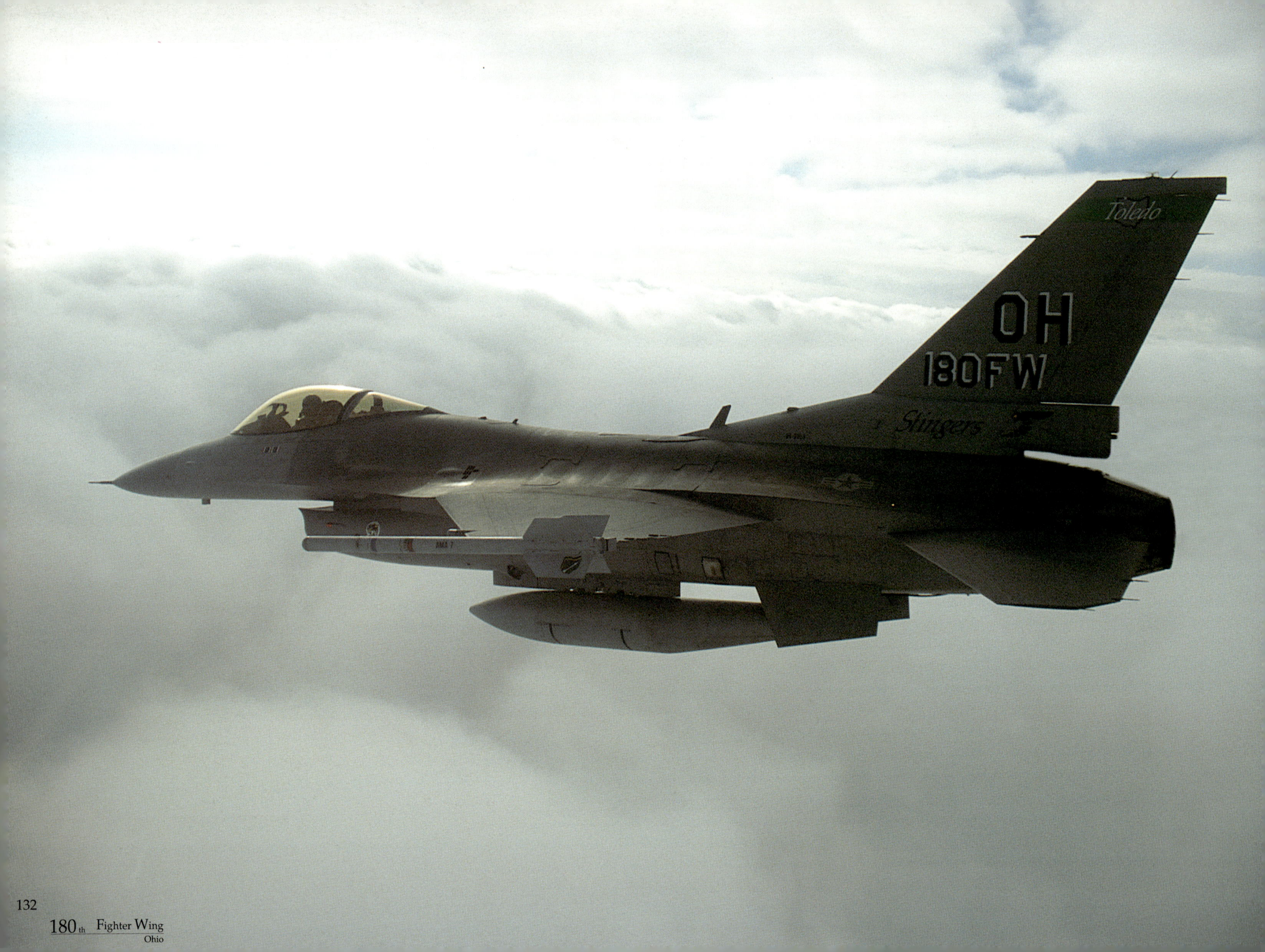
Toledo
OH
180FW
Stingers

192nd Fighter Wing
Virginia

192nd Fighter Wing, Virginia (emblem approved 14 October 1983). The emblem represents the unit's primary mission to fly and defend freedom and democracy. The background of blue and gray represent the sky and the old Virginia heritage of Confederate service. The two white flight symbols represent - the dual role of the federal and state missions. The lightning bolt indicates the speed, force, and ability to accomplish the flying mission. The horse and rider facing the blue sky with upraised sword represents readiness. It also indicates the strong foundation of granite on which the unit is built. The monument itself represents the City of Richmond, "City of Monuments".

149th Fighter Squadron, Virginia (emblem approved 1 March 1956). The eagle represents the national bird of the United States, the ultimate air power of the skies. The eagle carries two weapons of aerial combat which the squadron uses. On 12 March 1992, the Confederate "stars and bars" background was removed from the emblem by gubernatorial decree.

149th Fighter Squadron, Virginia.

Constituted in September 1942 and activated at Mitchell Field, New York as the 328th Fighter Squadron. Assigned to the 352nd Fighter Group, 8th Air Force, the 328th FS flew combat missions in the ETO with Republic P-47Ds beginning in September 1943. The squadron converted to P-51 D Mustangs in April 1944 and continued to fly combat missions until VE Day. The unit was allotted to the Virginia National Guard and redesignated as the 149th Fighter Squadron on 24 May 1946. The 149th Fighter Squadron (Single Engine) was re-equipped with P-47D Thunderbolts at Byrd Field, Sandston, Virginia, near Richmond, on 21 June 1947. On 1 March 1951, the squadron was mobilized for the Korean War and moved to Turner AFB, Georgia for service with Strategic Air Command. It transferred to Tactical Air Command as the 149th Fighter-Bomber Squadron and moved to Godman AFB, Kentucky. It returned to state control on 1 December 1952 at Byrd Field and was re-equipped with Douglas B-26B/C Invaders.

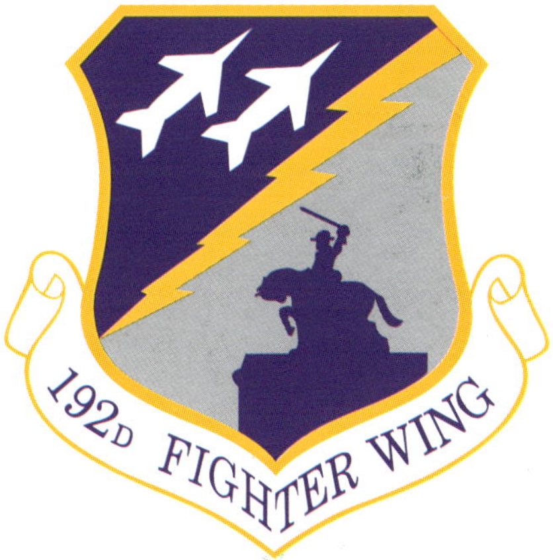

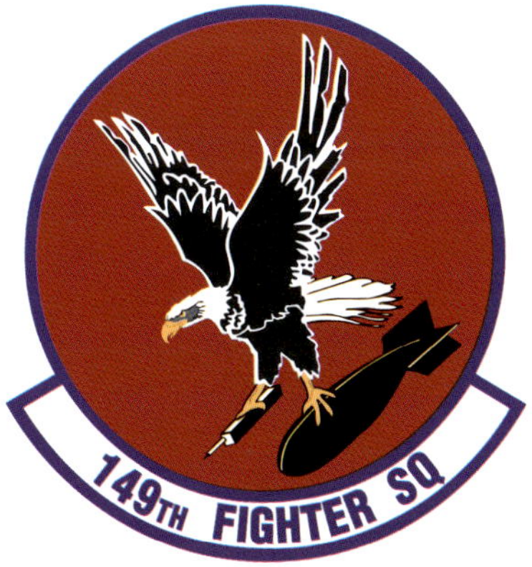

The unit was redesignated as the 149th Fighter Interceptor Squadron, but was redesignated again on 10 April 1958 as the 149th Tactical Reconnaissance Squadron. It converted to Martin RB-57 Canberras, but only one aircraft was received. Again on 14 June 1958 the unit redesignated as the 149th Fighter Interceptor Squadron and was converted to Republic F-84F Thunderstreaks.

On 1 October 1961, the squadron was mobilized for the Berlin Crisis, but remained at Byrd Field. The unit retumed to state control on 20 August 1962. The 192nd Tactical Fighter Group was activated on 15 October 1962 as the squadron headquarters.

The unit converted to Republic F-105D/F Thunderchiefs beginning on 19 February 1971. In 1982 the unit converted to Vought A-7D Corsairs and then to F-16C/Ds from 1991 to 1992. On 15 March 1992 the unit was redesignated as the 149th Fighter Squadron and on June was gained by Air Combat Command.

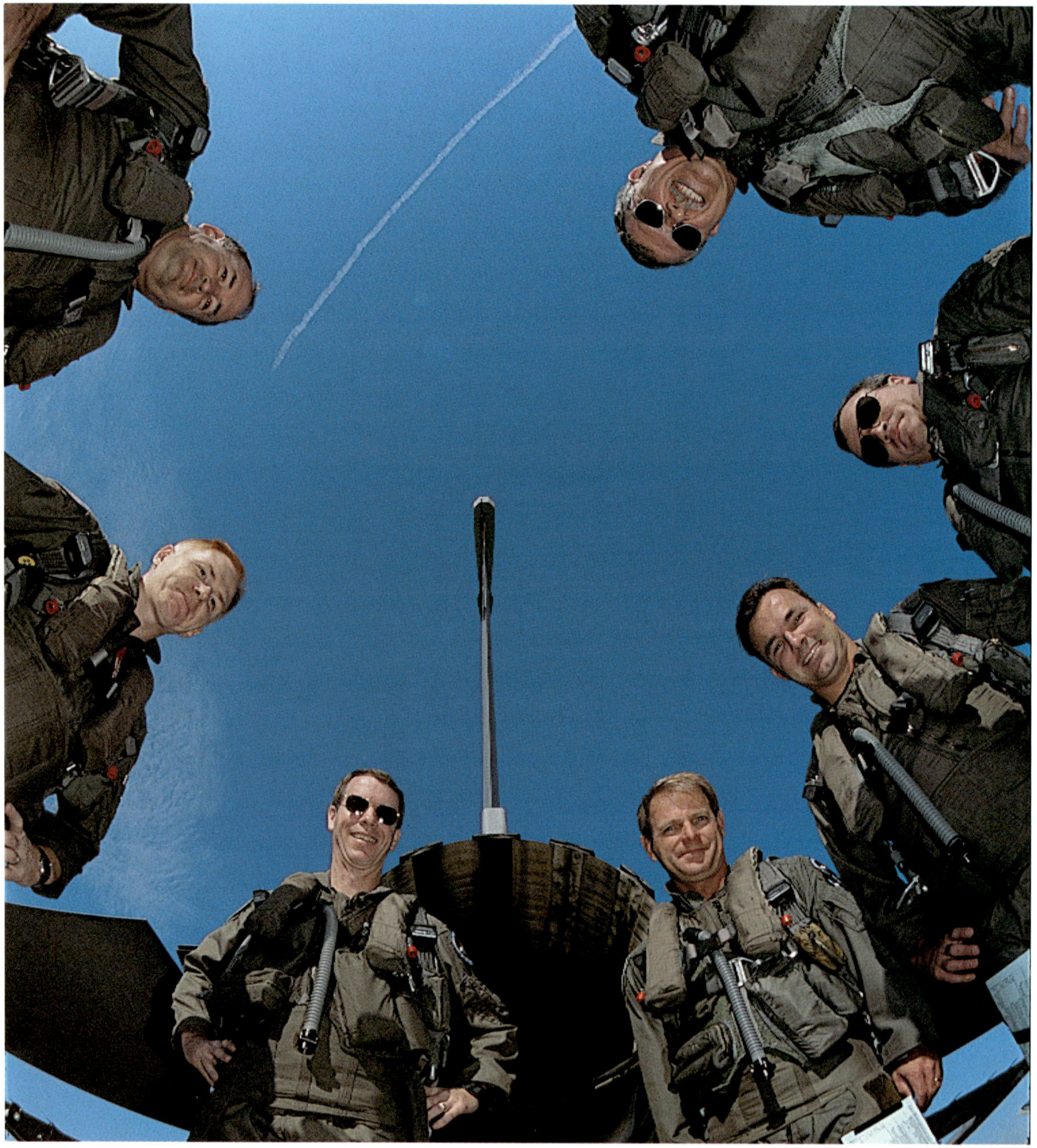

On the right
192nd Fighter Wing pilots pose behind the tail of an F-16C Block 30 Fighting Falcon.

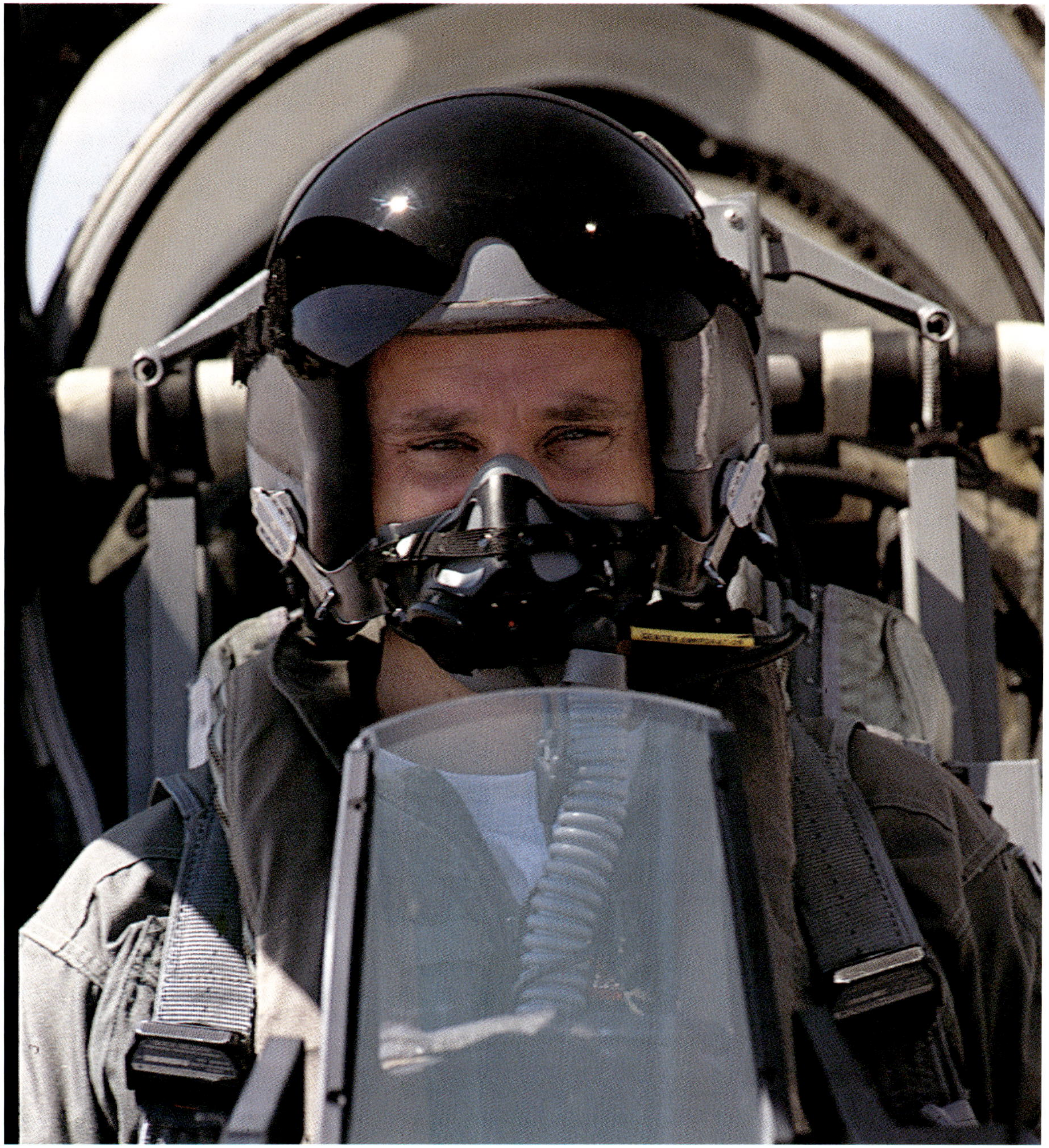

"GREAT CIRCLE" AT THE FIRING RANGE

BY GIAMPAOLO AGOSTINELLI

Ready for take-off, at 2:00 pm sharp, we join with four F-16Cs near the runway head. Lt.Col. Steve Hicks, who flies my F-16D, reminds me that, according to briefing, we are the last to go, after the four-ship formation will have released their brakes at 10 seconds interval.

"OK, we go!" advises me on the radio, and suddenly he selects afterburner thrust, swinging the throttle handle aside and pushing it full forward. The engine roars all its powerful thrust, pushing on my back.

My eye catches the fuel flow dial revving up wildly: 45...51...53...58...61,000 pounds per hour; the General Electric F110-GE-100 afterburning turbofan swallows and burns half a ton of kerosene in one minute. Once aloft we keep the AB on in order to reach the formation quicker, as we were the last to take-off.

On the left
A 192nd FW pilot is seen behind the semi-transparent Head-Up Display mirror.

Perfect rejoin: we take the left station of a neat four-ship "fingers" formation proceed enroute to a range south of Richmond, on the Virginia-North Carolina border. There we will release the small practice, blue-colored bombs and then we will do some nice gunnery with our 20-mm M-61 Vulcan six-barrel gun, with a couple of passes over a special electronic-scoring banner. In 15 minutes, with a right-hand break we space ourselves to begin the passes over the range half a mile down below.

We are the last in the long queue swinging in the sky, preparing for the dive followed by a 5.5-Gs pullup just after releasing the bomb.

"Gone! Do you see it? I'm banking a bit, so you should see it more easily." Steve says. I'm able to track it for some seconds, then I lose it in the colored background. The "great circle" goes, and we dive and dive again to the target.

In these pages
The 149th FS F-16C Block 30 fighter-bombers are also capable to accomplish photo-reconnaissance missions.

Below
The landing - nosewheel up - of an 149th FS F-16C.

And, on the left
Another four-ship is ready for their turn at the firing and bombing range.

In the page on the right
A Viper as it is seen in close formation under a cloud layer.

In the page overleaf
A four-ship "fingers"formation of 192th FW F-16Cs as it is seen from the rear cockpit of the leading F-16D.

In the next page
Echelon turn of four F-16Cs before their run-in onto their target at a firing range.

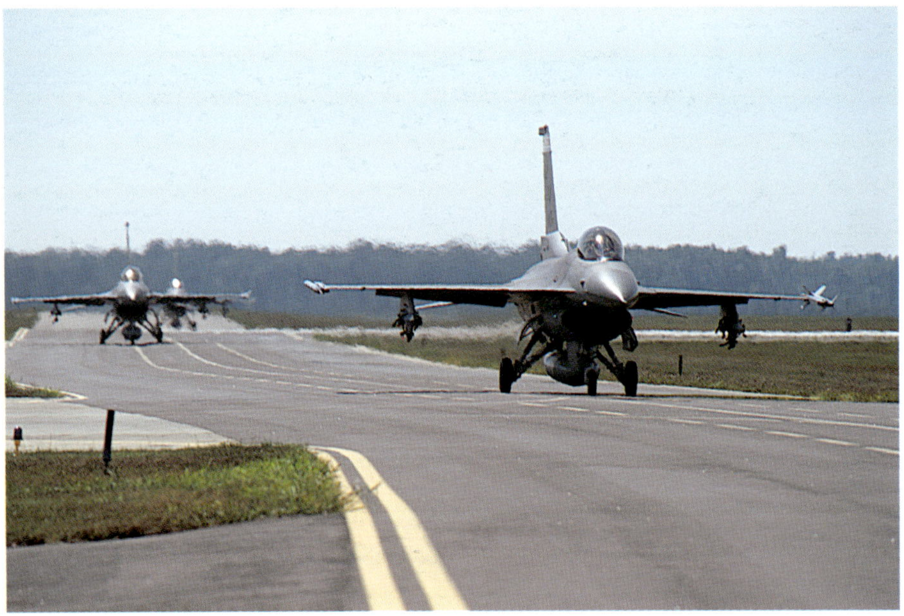

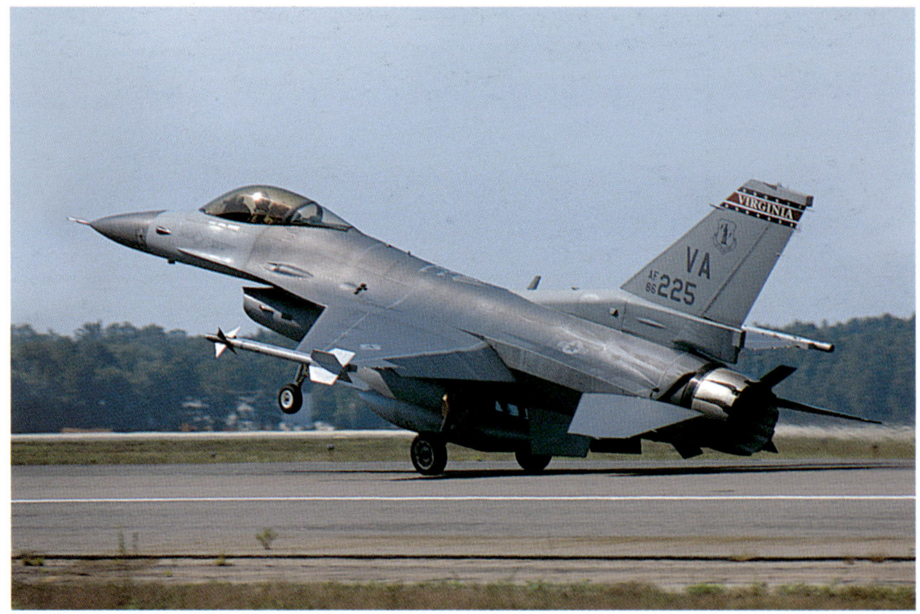

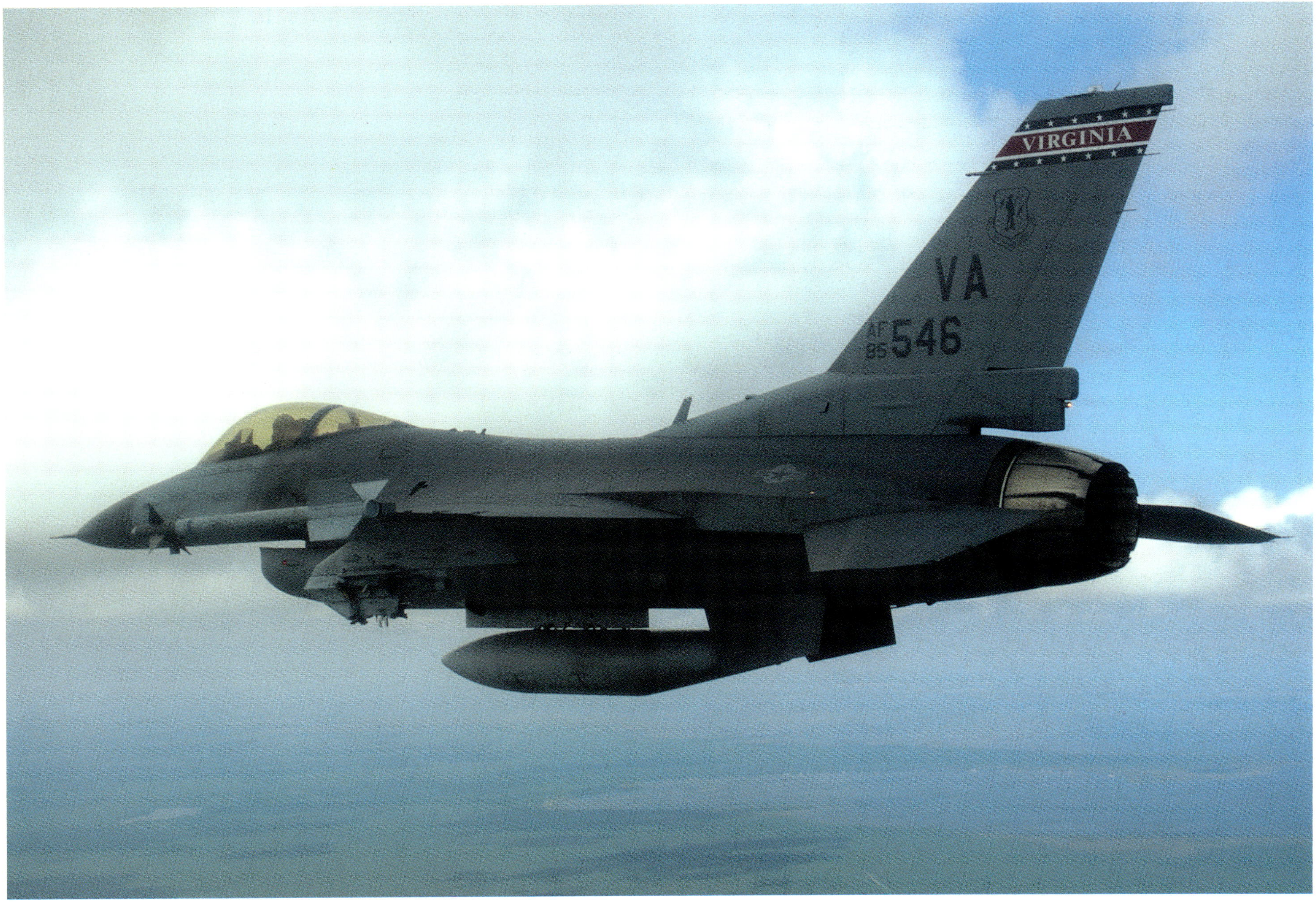
VIRGINIA
VA
AF
85 546

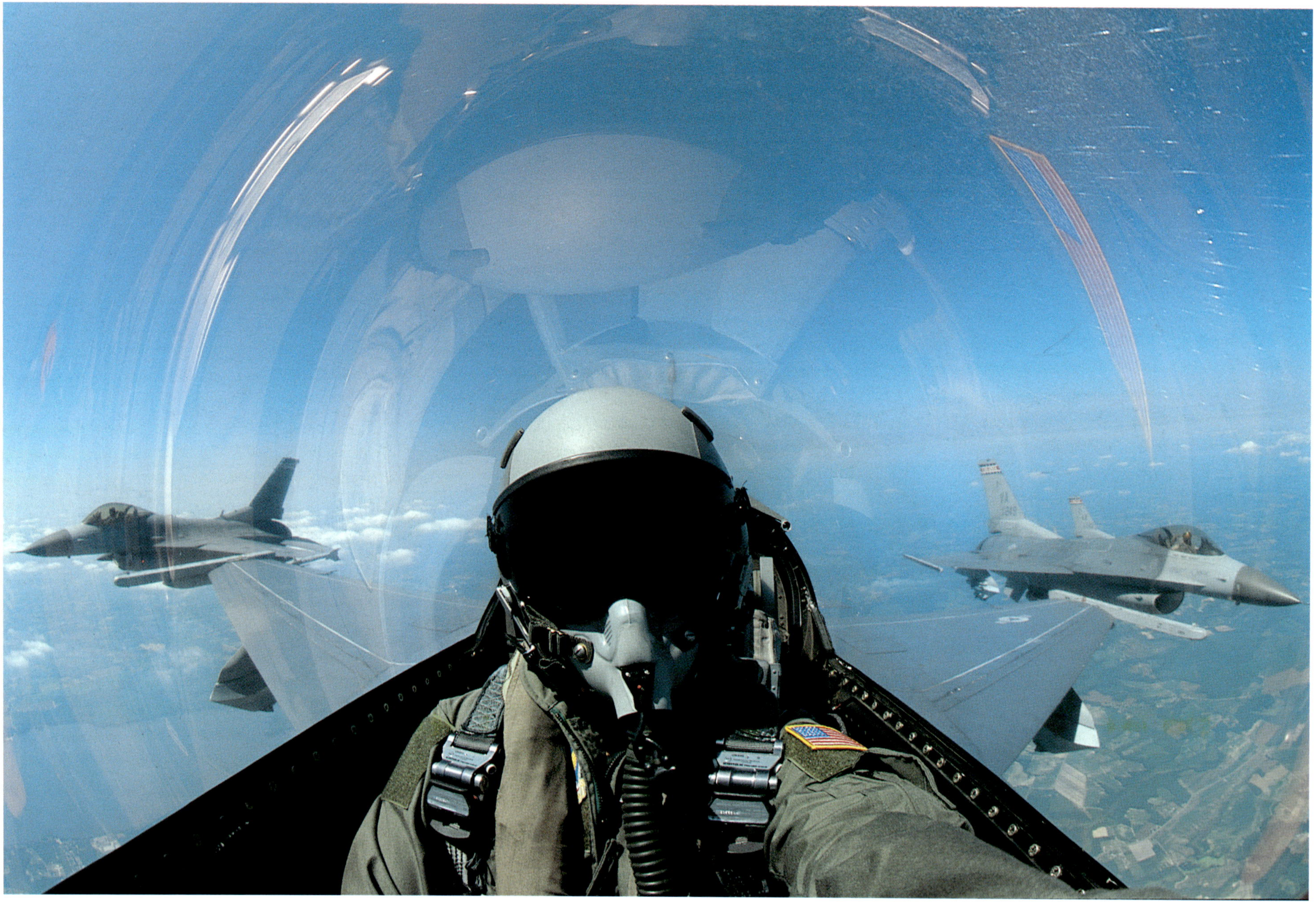

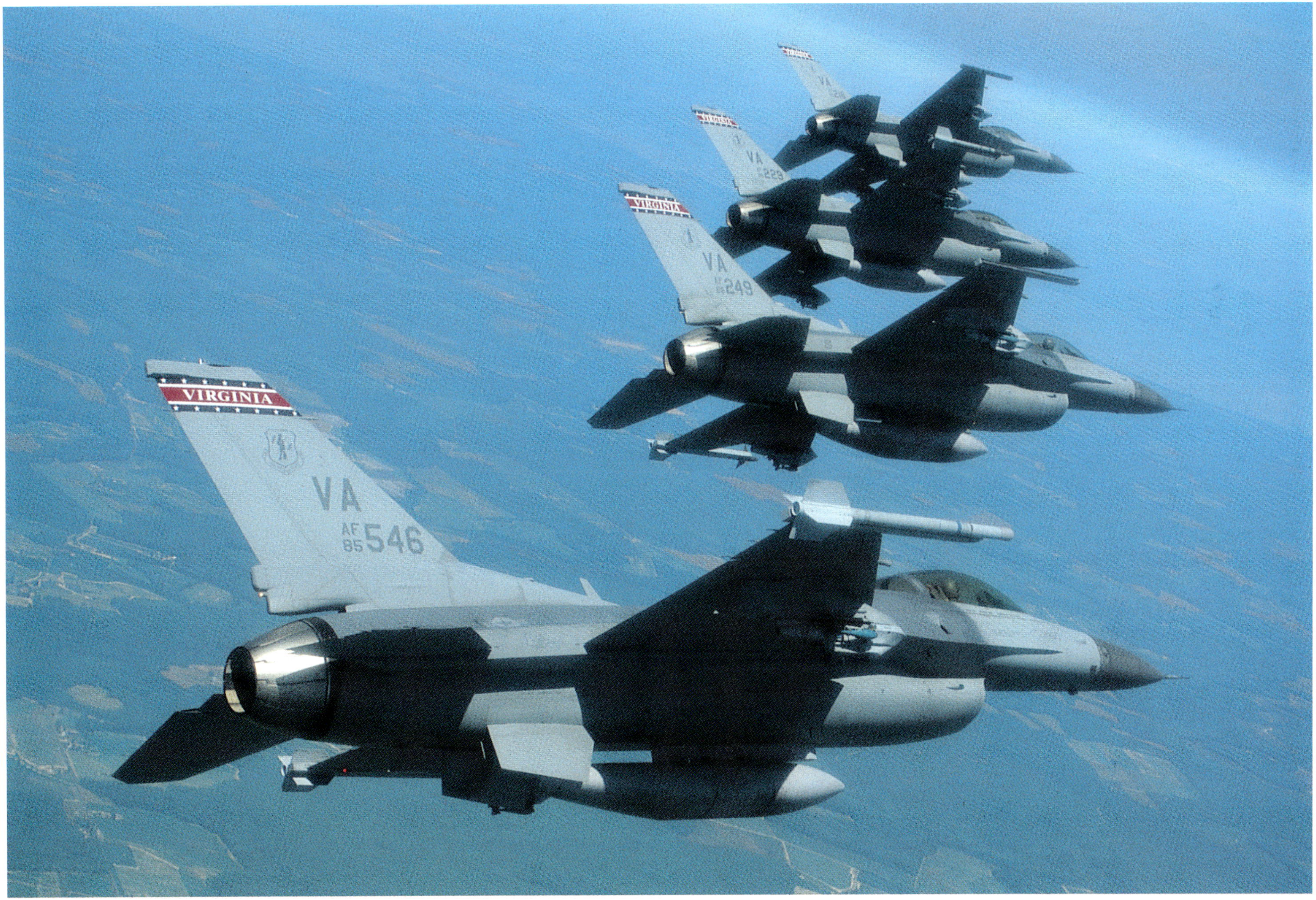
VIRGINIA
VA
AF
85 546
VIRGINIA
VA
249
VIRGINIA
VA
229

Now it's for the gun: Steve starts a gentle climb, then banks the plane to inverted and pulls it in a steep dive. "Inverted, we see better downwards" he says while we point towards the banner. "Ready?". My affirmative answer is overwhelmed by the Viper's shiver and the barking of a ripple of 20-mm rounds shot by the M-61 Vulcan gun.

A quick pullup followed by a 90 degrees right turn simulates a tactical egress profile to escape from triple A.

5.6 Gs make themselves felt, but we are out shortly. Then we get back in line with the others for a second gun pass with the same procedure. Again upside down, we start our vertiginous dive to the target.

A brief ripple, and we head away immediately to rejoin the formation.

"Now we are going to split the formation. We and Number Two are simulating a tactical recce mission" Steve advises. Flying a little apart we lose altitude to simulate a low-level approach to the target that we are to record on film - a bridge which joins the shore line with an island in farthest North of North Carolina.

In tactical formation we approach the first turn point, another bridge on a river, and then we set a north-easterly heading of some 45 degrees.

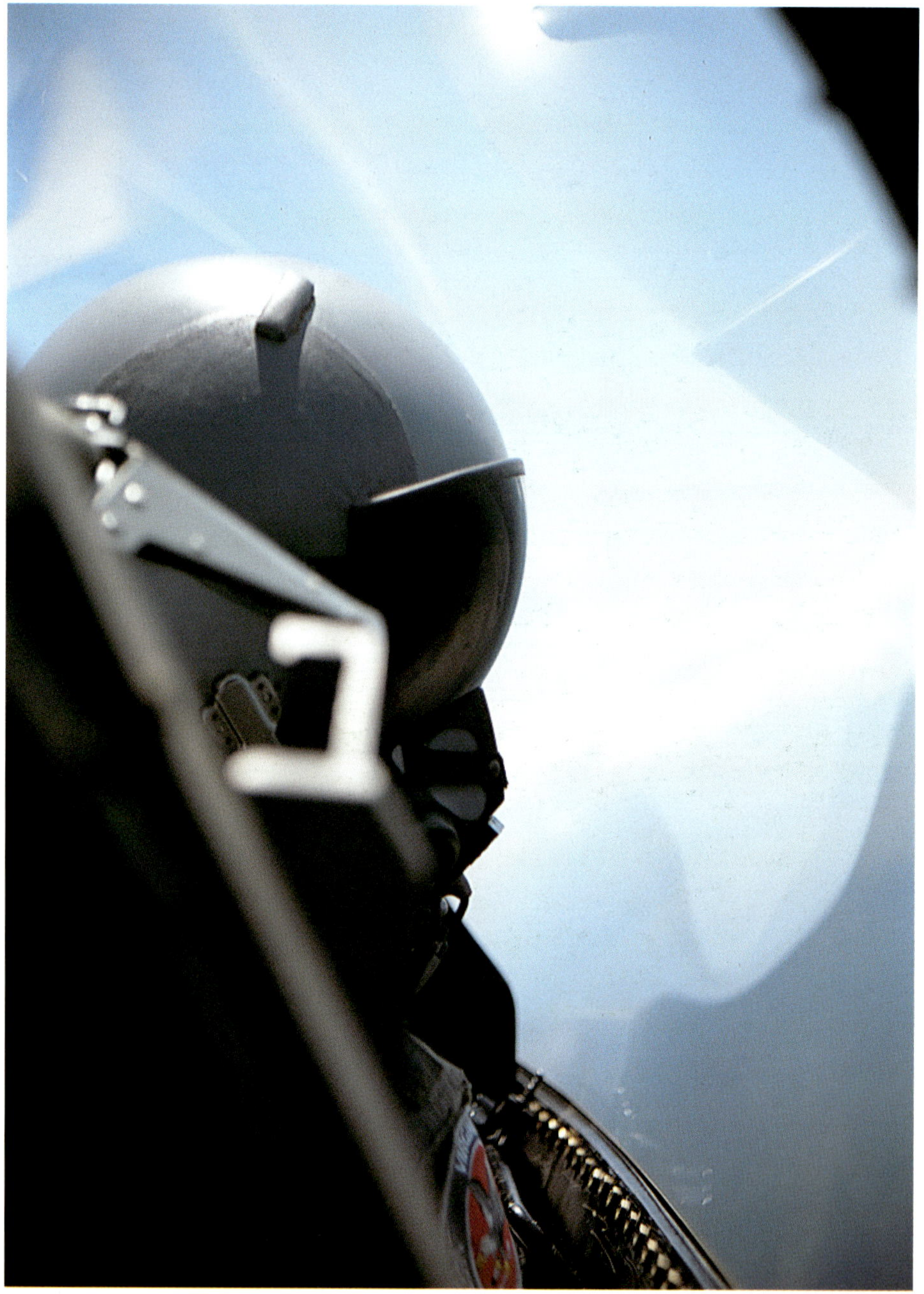

On the right
Lt. Col. Steve Hicks with his helmet visor lowered, seen from the rear cockpit of an F-16D.

In the page on the right
A view of an F-16D rear cockpit and on an F-16C in close formation.

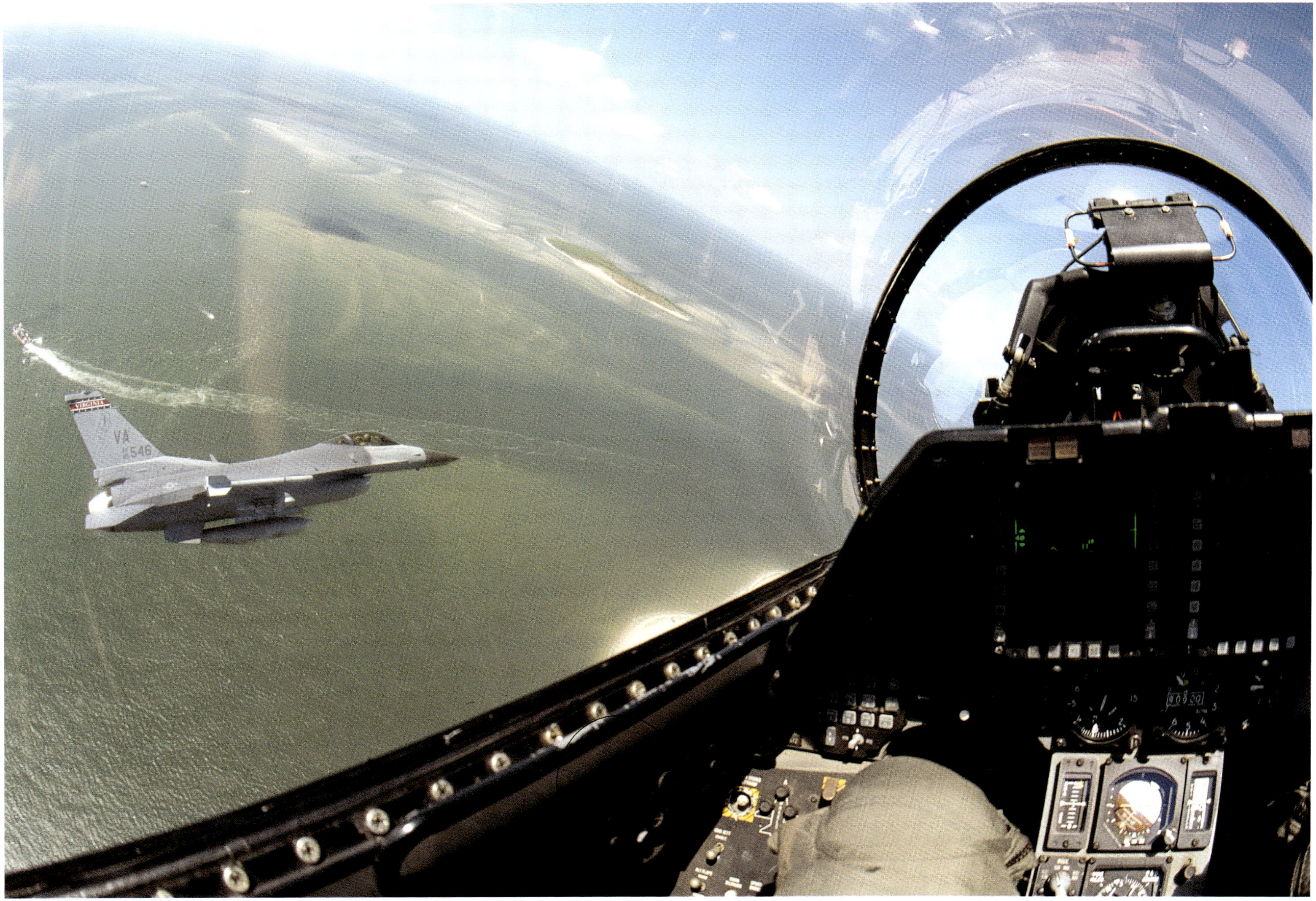
VIRGINIA
VA
546

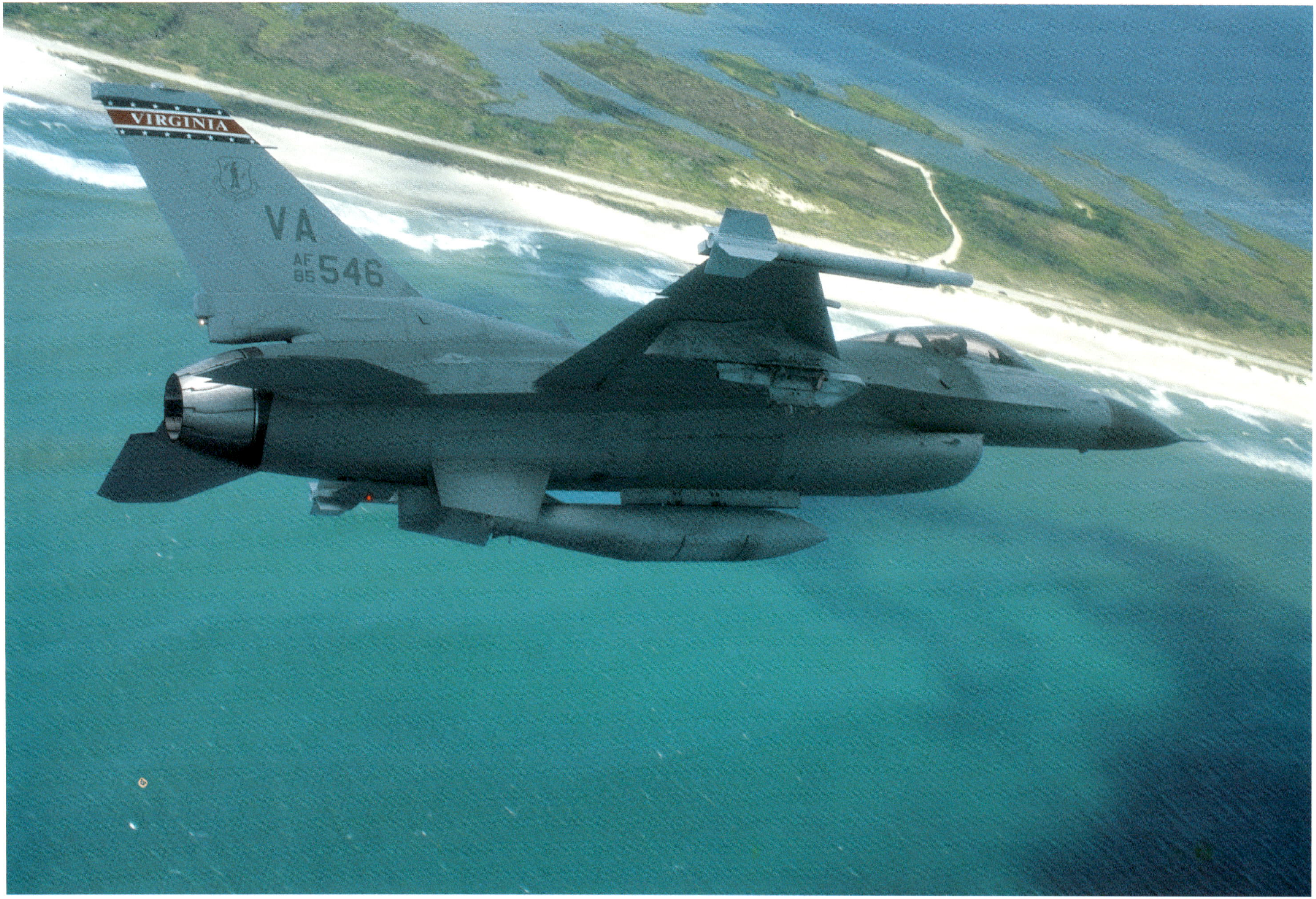
VIRGINIA
VA
AF
85 546

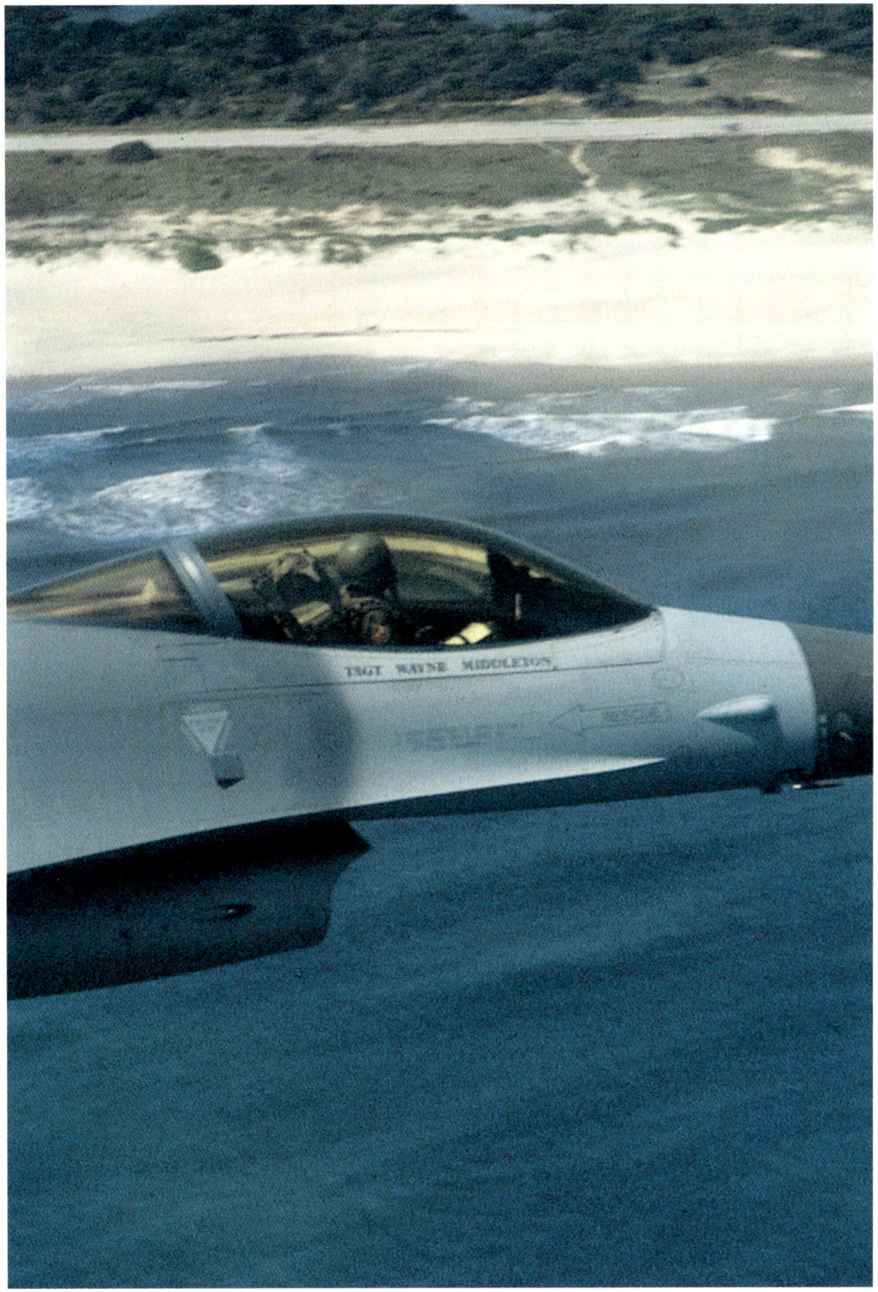

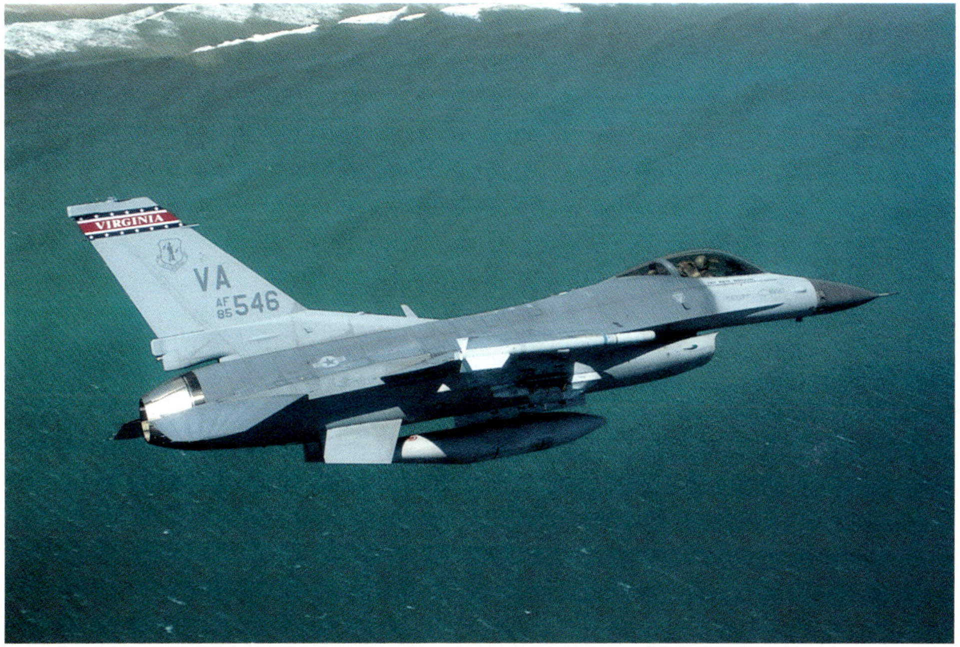

Above
One more approach turn onto the firing range.

On the left
"Closing" on the cockpit of an F-16C Block 30 of the 149th Figther Squadron.

In the page on the left
A Viper flying along the Virginia oceanic coast.

Speed is Mach .80: my mission computer shows 4 mins, 30 secs to target on the right-hand multi-function display, whereas Steve can see everything through his Head-Up Display without need to lower his sight inside the cockpit and distract himself from handling the aircraft, low "on the deck". Minutes fly away fast, we are tho bolts hurled to the target. After a wide left turn we circle the bridge on the right in order to get it with a long film strip.

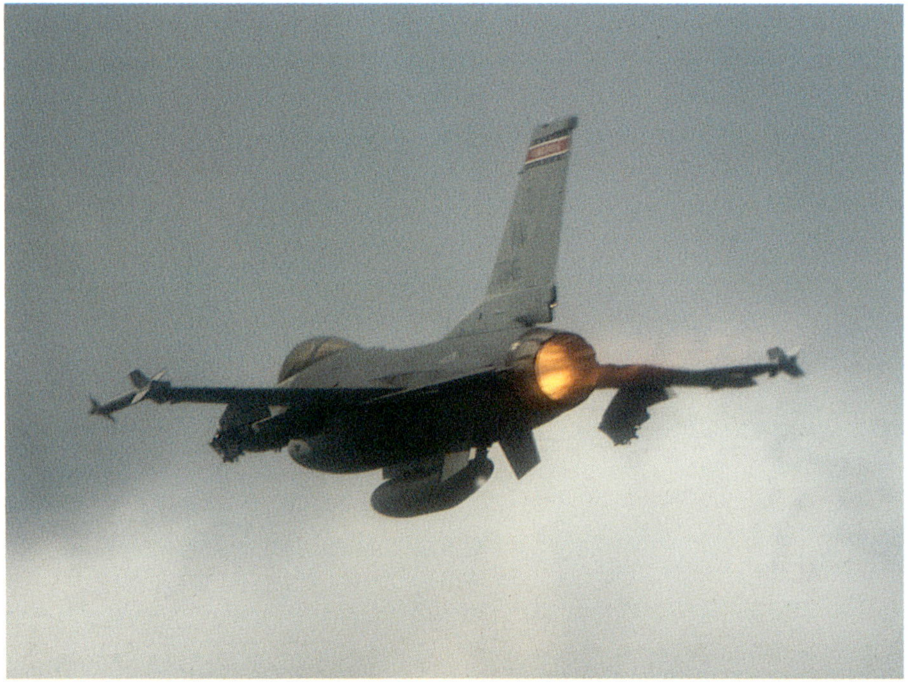

Above and in the page on the right
An F 16C Block 30 accelerates for its run-in under the augmented thrust of its General Electric F110-GE-10 afterburning turbofan engine.

In the page overleaf
Mission accomplished. A Virginia ANG Viper turns on the Ocean during its approach to Byrd Field

We accelerate to Mach .93 at 500 feet above ground level, and Steve starts "shooting", simulating to operate the control panel of the big KS-87 camera. The GPS - included in the recent upgradings to the 192th Fighter Wing Vipers to better exploit the special photo-recce pod - is of great help to verify the target coordinates and then our own position precisely.

"Using the big photo-recce pod's left-hand window, and with an ample turn radius, we keep slightly out of the hottest area, and in the same time we make sure to shoot much more film in a single pass" Steve explains.

"In real combat missions we reconnaissance pilots have a very difficult and dangerous role to fulfil, because we are called mainly to assess the attack damages, therefore we have to enter a very 'warmed-up' area, sometimes flying in a complex orographic situation. Even nowadays one cannot rely on far-away though sophisticated spy-satellites, therefore we must be always prepared to accomplish, in the best possible way, the traditional, and perhaps low-tech, photo-reconnaissance missions, because they are likely to achieve the best results, in certain conditions."

Then we head back home. We have been some 50 minutes aloft and fuel is sensibly diminished from 10,400 lbs on leaving blocks, to some 2,400. We have lit up the afterburner several times and flown a long stretch at low level, where fuel consumption is higher, and the maneuvres at the range had their own share.

Richmond International Airport is in sight. Breaking to the left over the runway, we space ourselves for landing. I meet Steve under the ladder, on the tarmac of the 192nd Fighter Wing flight line. We congratulate each other warmly.

"As you have been able to see, we are the 'Jacks-of-all-trades' of the Air Guard. But this is gratifying for us, a qualification we are proud of."

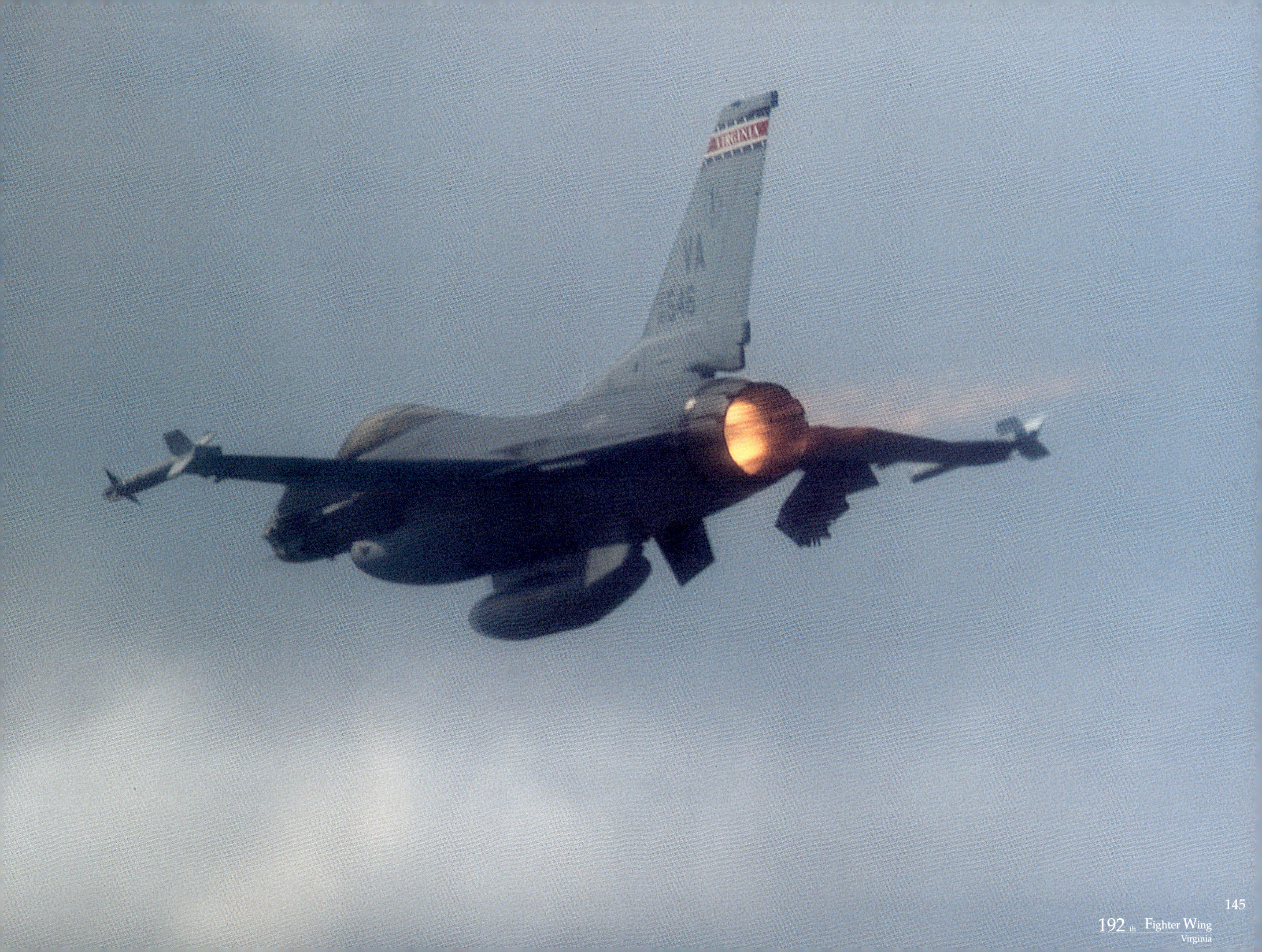
VIRGINIA
VA

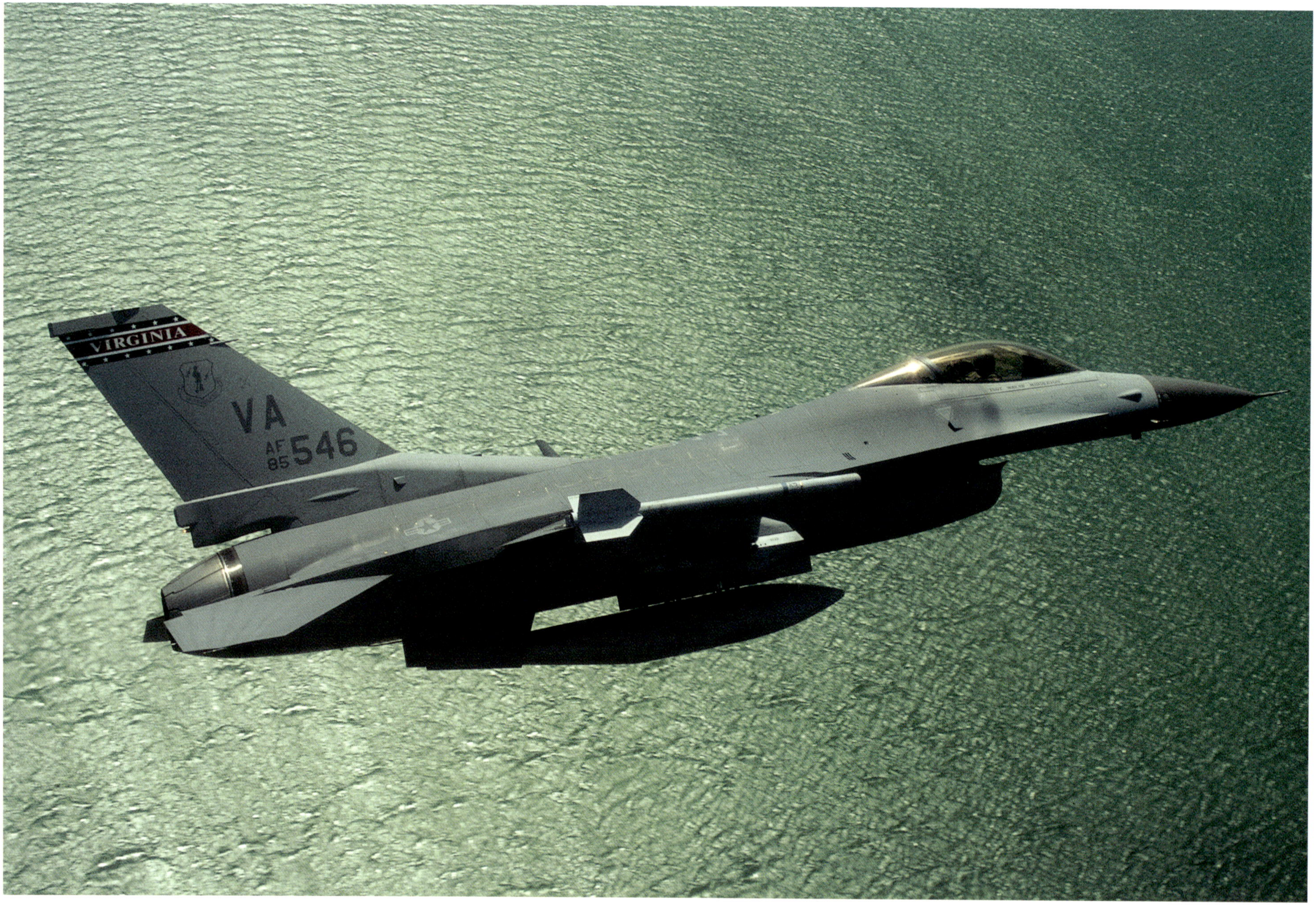
VIRGINIA
VA
AF
85
546

31st Fighter Wing
Aviano

31st Fighter Wing, Aviano Air Force Base, Italy

The mission of the 31st Fighter Wing is to conduct and support air operations in Europe's southern region and to maintain munitions for the North Atlantic Treaty Organization and national authorities. The 31st FW maintains two F-16 fighter squadrons, the 555th and the 510th, capable of conducting offensive and defensive air combat operations under NATO, Supreme Allied Commander Europe and national taskings utilizing various weapon systems. In peacetime, the 31st FW prepares for this role by maintaining its aircraft and personnel in a high state of readiness. The 31st FW also includes the 603rd Air Control Squadron, capable of providing air surveillance, control and communications.

The 31st FW is the only U.S. fighter wing south of the Alps. During a NATO crisis, the wing's operational forces become part of the 5th Allied Tactical Air Force, located at Vicenza, Italy. This, and its strategic location, make the wing critical to operations in NATO's southern region.

Beginning July 1994, the wing provided combat support for NATO's first-ever operational mission, *Deny Flight*. In August and September 1995, 31st FW F-16s flew more than 400 combat sorties during Operation *Deliberate Force*, paving the way for implementation of the Dayton Peace Accord. With the formal signing of this accord, Operation *Deny Flight* ended and the wing's emphasis shifted to support what is now Operation *Deliberate Forge*.

In March 1999, in support of Operation *Allied Force*, U.S. and Allied forces assembled at Aviano Air Base, Italy, to react if called upon by NATO leadership. More than 4,000 temporarily assigned personnel were brought to the base bringing the military population to over 7,000. During the 79-day operation, 8,527 missions were flown out of Aviano.

The 31st FW also supports four geographically separated units: 31st Munitions Squadron, Camp Darby, Italy; 31st Munitions Support Squadron, Ghedi AB, Italy; 31st "Red Horse" Flight, Camp Darby, Italy; and the 496th Air Base Squadron, Moron AB, Spain.

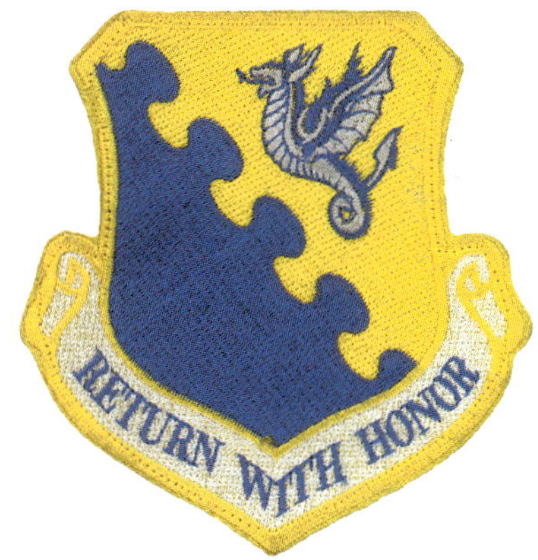

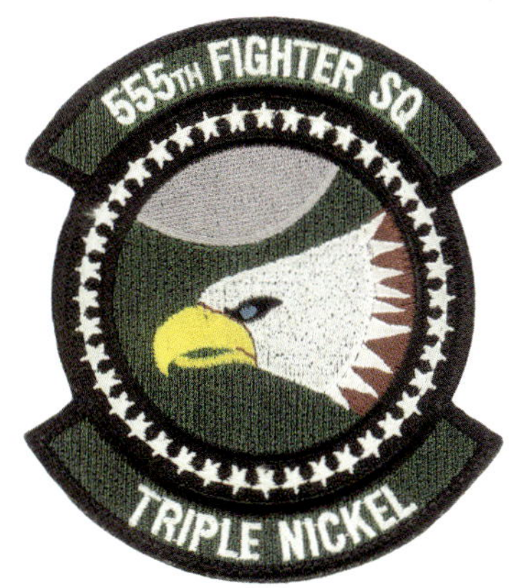

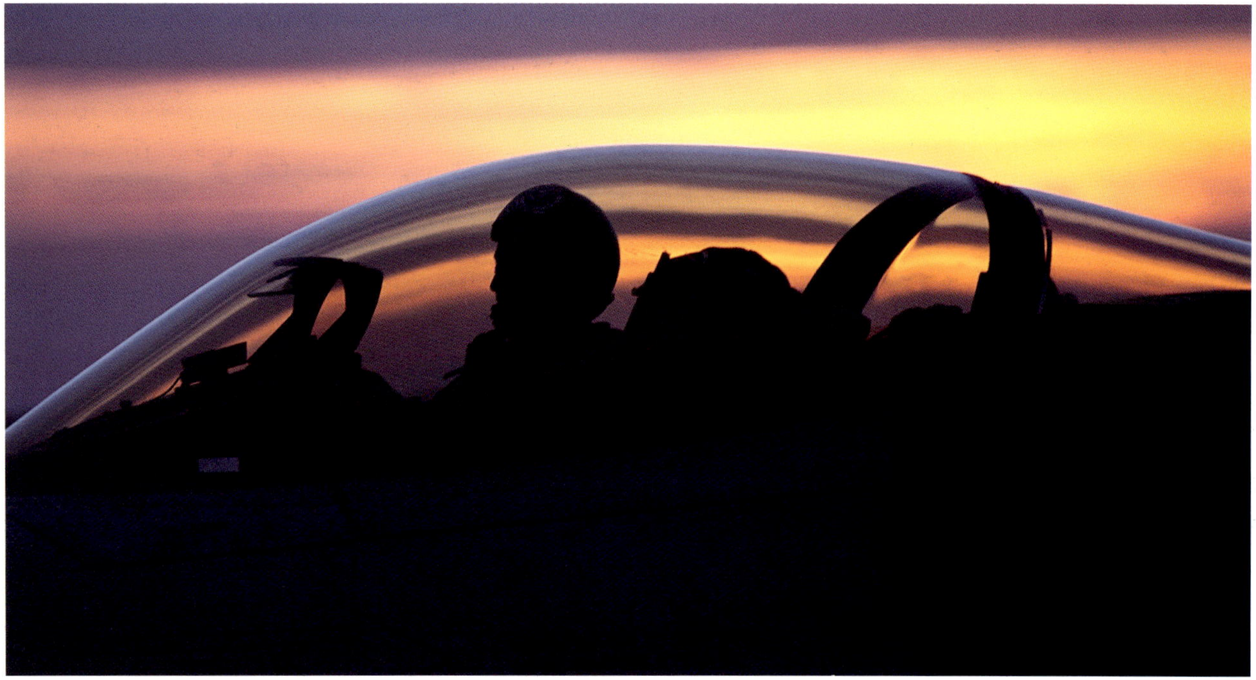

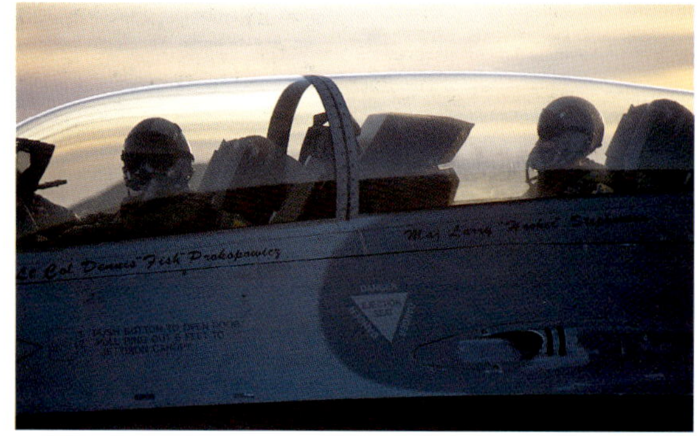

Above
The crew of the single 555th FS two-seat F-16D.

On the left
A "Triple Nickel" pilot in his Viper, taxiing at sunset.

In the page on the right
The F-16D that sports 555 as the last digits of its "buzz number" taxies at Aviano on its re-entry from a mission over the Adriatic Sea.

555th Fighter Squadron "Triple Nickel"

The 55th Fighter Squadron is one of two F-16 squadrons of the 31st Fighter Wing. The 555th, better known as the "Triple Nickel", was originally constituted as the 555th Bombardment Squadron in 1942. The unit was stationed at locations in England, France, and Belgium during World War Two, flying the Martin B-26 Marauder and the Douglas A-26 Invader. It was inactivated after the war, but brought back as the 555th Tactical Fighter Squadron in 1964, flying the McDonnell F-4 Phantom fighter. The unit saw combat in Southeast Asia, from February 1966 until the close of hostilities. The Triple Nickel became known as the "World's Largest Distributor of MiG Parts" by downing more enemy MiG fighters than any other squadron from any branch of the U.S. military involved in the war. The 555th's current green-and-white squadron emblem is derived from its Vietnam-era insignia.

Redesignated as the 555th Tactical Fighter Training Squadron in 1974, the unit relocated to Luke AFB, Arizona and trained new pilots in the McDonnell Douglas F-15 Eagle. It deactivated in March 1994 shortly before reactivating a month later at its present location in Italy, where it transitioned to the then General Dynamics and now Lockheed Martin F-16 Block 40 Fighting Falcon. The 555th's immediate mission was to contain the civil war in Yugoslavia. One month after arriving at Aviano AB, the Triple Nickel began flying contingency operations in Bosnia-Herzegovina. The unit started operations in support of

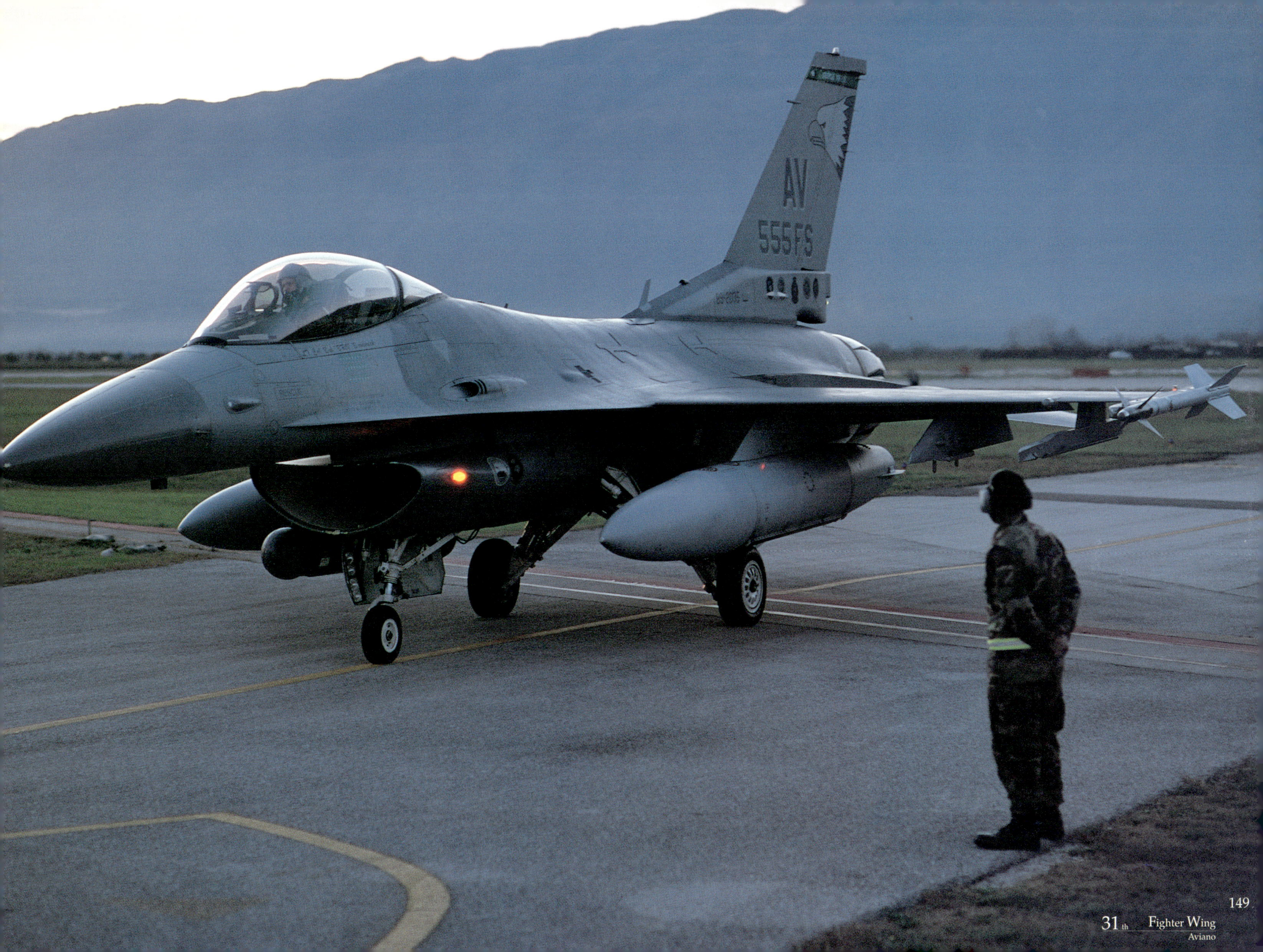
AV
555FS

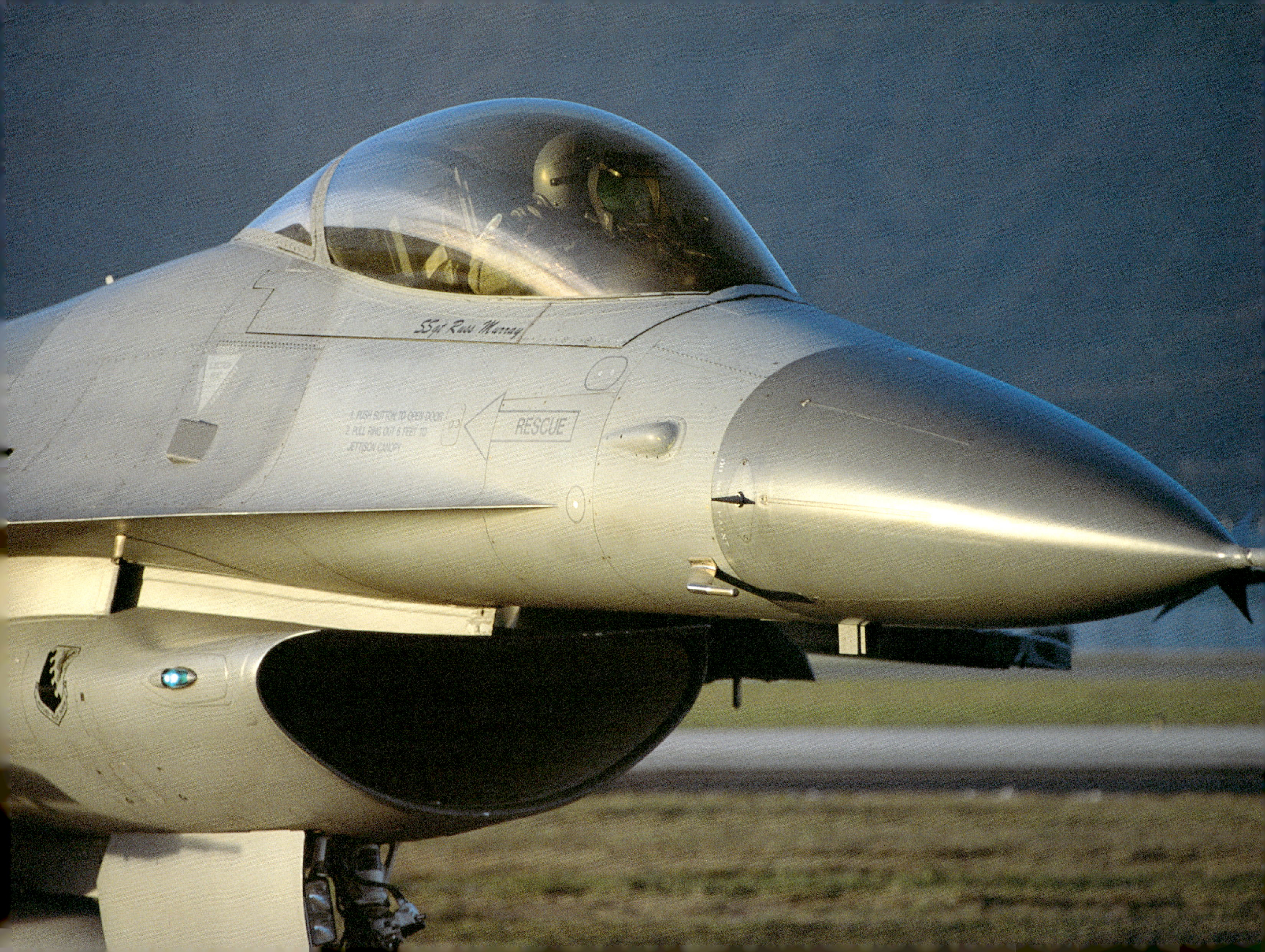
SSgt Russ Murray
1. PUSH BUTTON TO OPEN DOOR
2. PULL RING OUT 6 FEET TO
JETTISON CANOPY
RESCUE

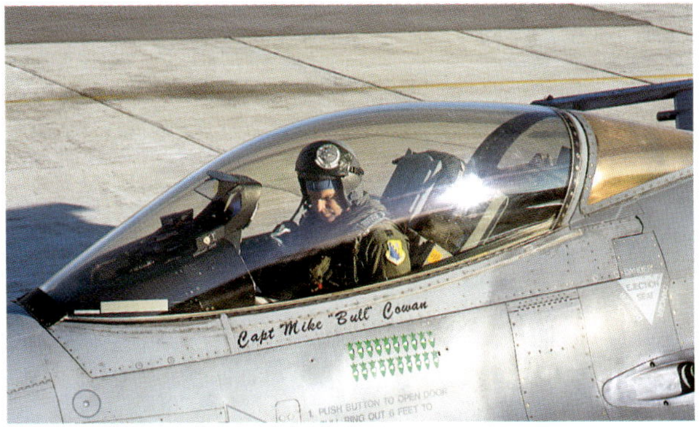

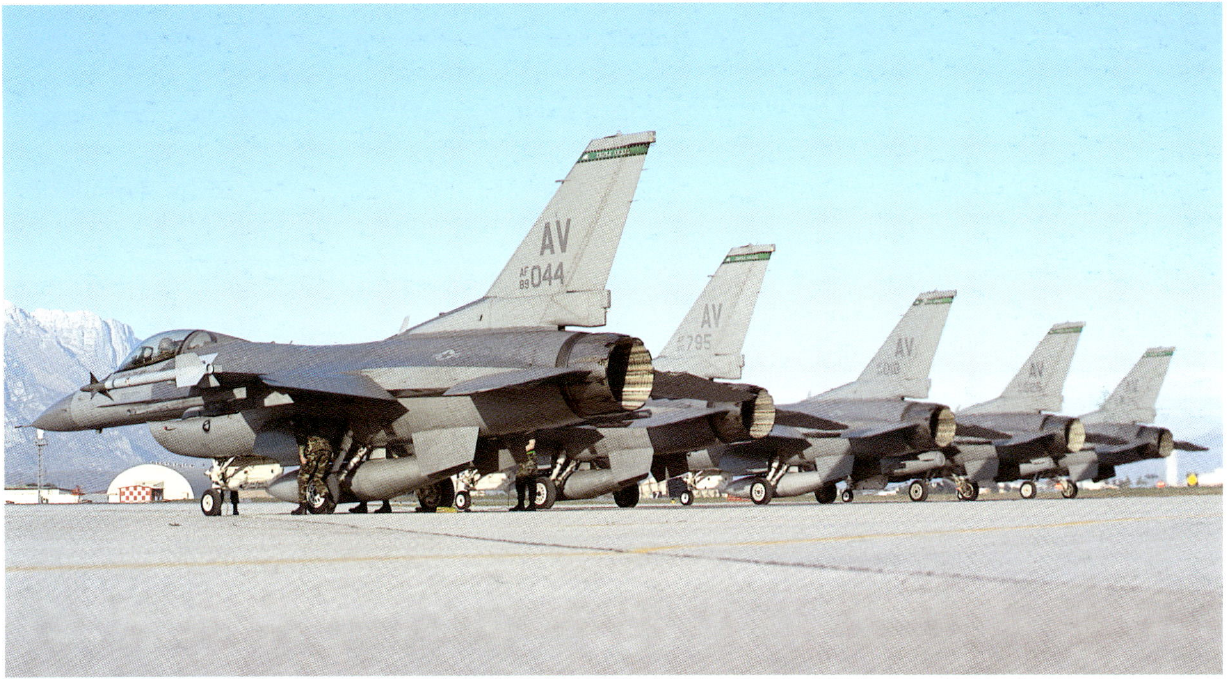

Above
A 555th FS pilot eases himself in the cockpit of an F-16C.

On the right
Five F-16s hold position near the main runway at Aviano AFB.

On the left
The sunset light makes this Viper glare golden.

In the page overleaf
Two Vipers in flight over the Dolomites.

Operation *Deny Flight*, and by September of the same year, the 510th joined her sister squadron. Both squadrons conducted daily sorties with other NATO forces, enforcing the "No Fly Zone" over the region. By 28 August 1995, the wing had flown 1,644 sorties in support of Operation *Deny Flight*. The 31st Fighter Wing led the largest air raids in Europe since World War Two in the summer of 1995 in Operation *Deliberate Force*, the NATO-led bombing offensive. *Deny Flight* resumed from them until December 1995 when the United Nations turned over authority for the security of Bosnia-Herzegovina to the NATO Implementation Force. F-16s from the 510th and the 555th squadrons at Aviano continued to maintain the security of Bosnia-Herzegovina under various operations including *Joint Endeavor*, *Deliberate Guard*, and *Deliberate Forge*, and *Decisive Edge*. From March to June 1999, the 31 AEW at Aviano, flew in Operation *Allied Force*.

The largest expeditionary wing in air force history flew nearly 9,000 combat sorties and accumulated almost 40,000 hours of combat service over the skies of Kosovo, Serbia, and throughout the Balkan region. With zero combat losses, the wing accomplished much during *Allied Force* as the two permanently assigned flying squadrons, the 510th and 555th, combined for nearly 2,400 sorties over 10,000 combat hours. The 555th reached the 1,000 combat sortie mark in May 1999. Since then, the squadron has deployed to Southwest Asia in support of Operation *Southern Watch*, enforcing the No-Fly Zone South in Southern Iraq. What follows are some first-hand accounts from various F-16 pilots involved in Operation *Allied Force*.

WORKING WITH F 16s BLOCK 50

BY ERIC HEHS

The HARM carried by the F-16 Block 50 in the SEAD mission (Suppression of Enemy Air Defense) doesn't have a big warhead. A HARM has to guide on the radar signal. Most SAM operators that are smart are not going to leave that radar on very long. That's good for us: if that radar isn't on, we're not going to get hit. But the HARM is not going to guide on the site either. Even if a HARM hits a site, it may blow up a radar dish or put a couple of holes in it. In most cases, though, it won't put a SAM site down for the count. On the other hand, a 2,000-pound bomb will and so will a Maverick.

Block 50 pilots perform hunter-killer, or DEAD (Destruction of Enemy Air Defense, pronounced deed), missions themselves with the Maverick missiles. They can't drop laser-guided bombs.

We used the datalink capability between the Block 40 and the Block 50 F-16s to address this issue. The HTS pod on the Block 50 locates a SAM. Those coordinates are then transmitted to a Block 40, which carries a laser-guided bomb. The Block 40 would then target and destroy the SAM site.

We had some success with this tactic, but we never used it much because we had competing priorities that took precedence. We did have one big success when we teamed Block 40 and Block 50 on an early warning radar in Montenegro.

This particular early warning radar was a problem through most of the war. When we got a tip where it was located, we had an F-16 Block 50 go in and shoot a HARM at it. The HARM hit real close and did some damage. Then a pilot in a Block 40 used his targeting pod to drop an LGB on the site to completely destroy it.

The F-16 has been called the workhorse of the war, and the numbers support that contention. F-16 squadrons led the first two missions of the war and most of the other missions throughout the air campaign. We had LANTIRN, NVIS (night vision imaging system), and ALE-50 (towed decoy) because of foresight by our leadership. Those decisions paid off in *Deliberate Force* a few years ago and certainly paid off again during *Allied Force*.

We learned a lot about the use of the targeting pod, an incredible tool. We didn't use the pod that much differently, but we really wrung it out. We used it on every sortie for a number of things: setting up basic tanker rejoins, identifying ground targets, and lazing ground targets. We capitalized on its capability as a reconnaissance tool as well - we have one hell of a reconnaissance capability with the LANTIRN pod. We regularly imaged targets and got BDA with it throughout the war.

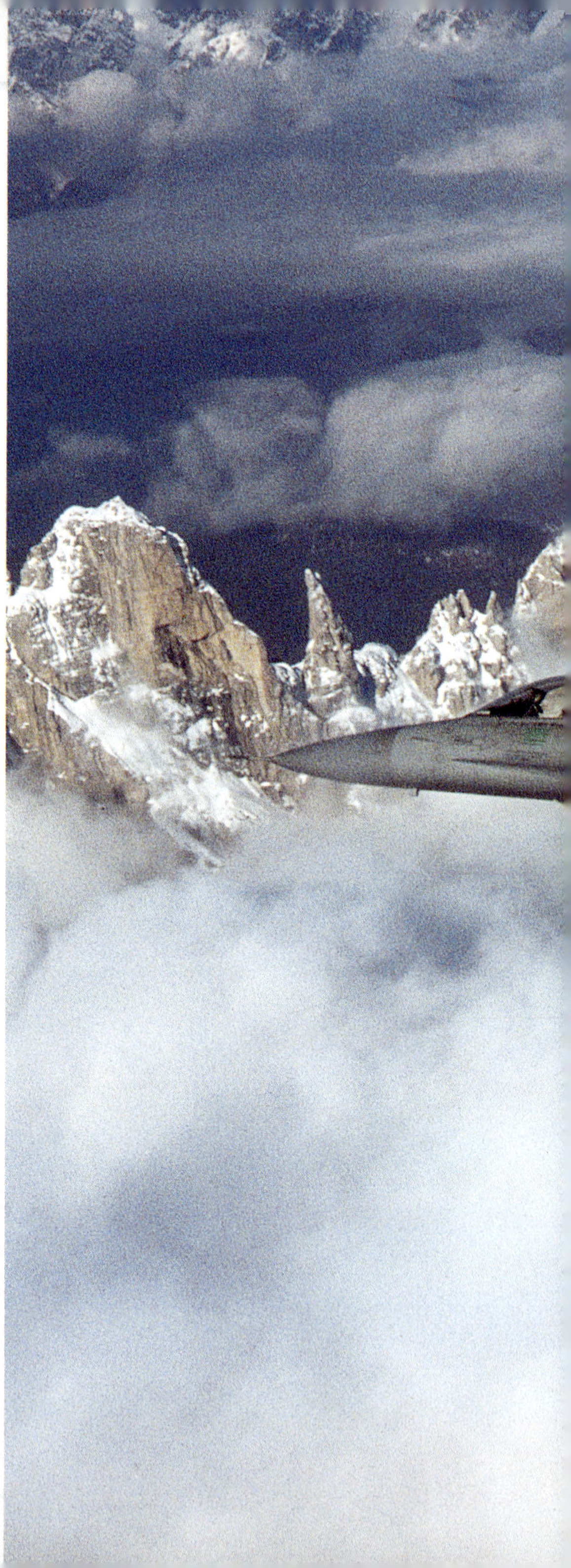

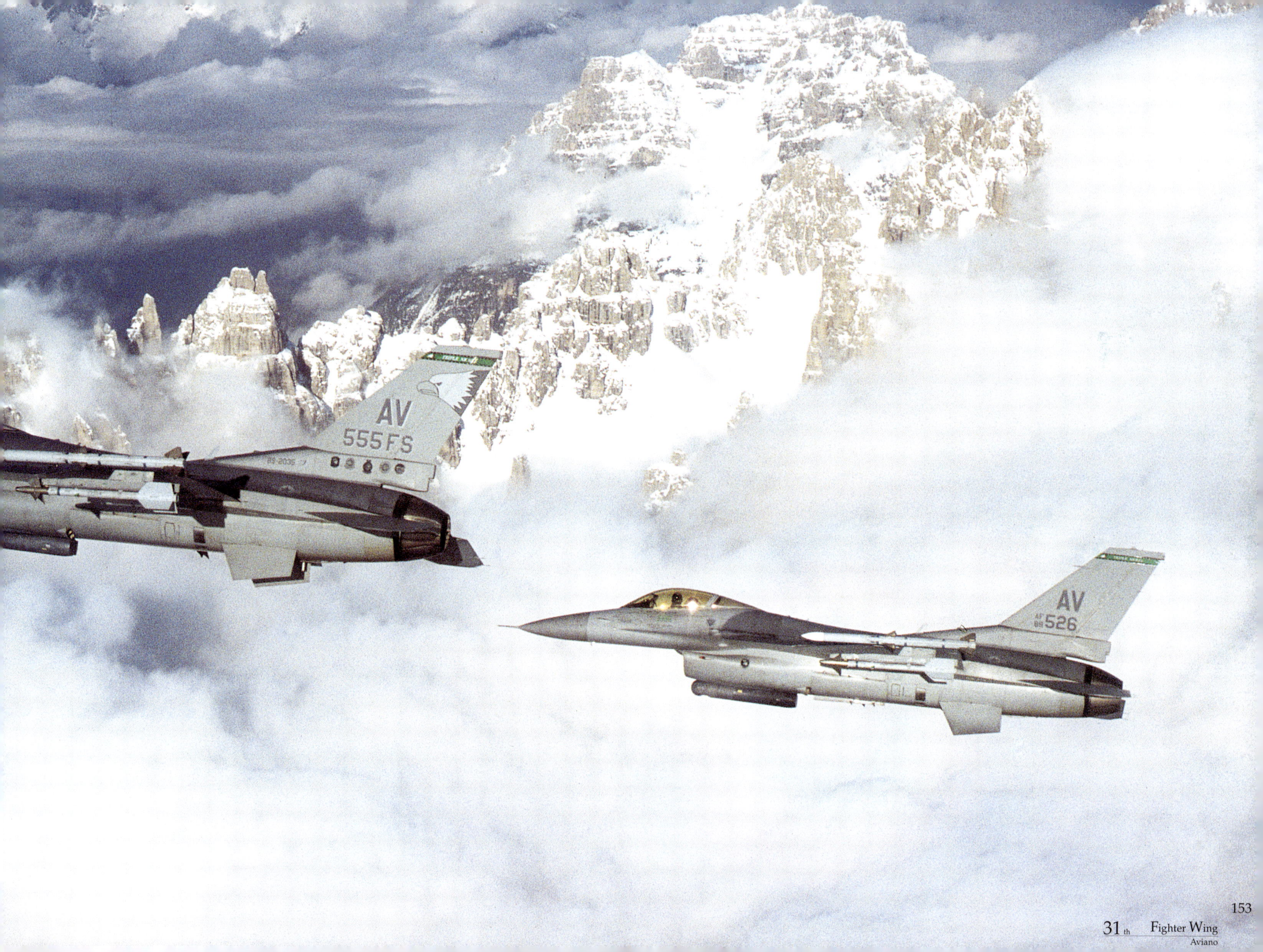
AV
555FS
AV
526

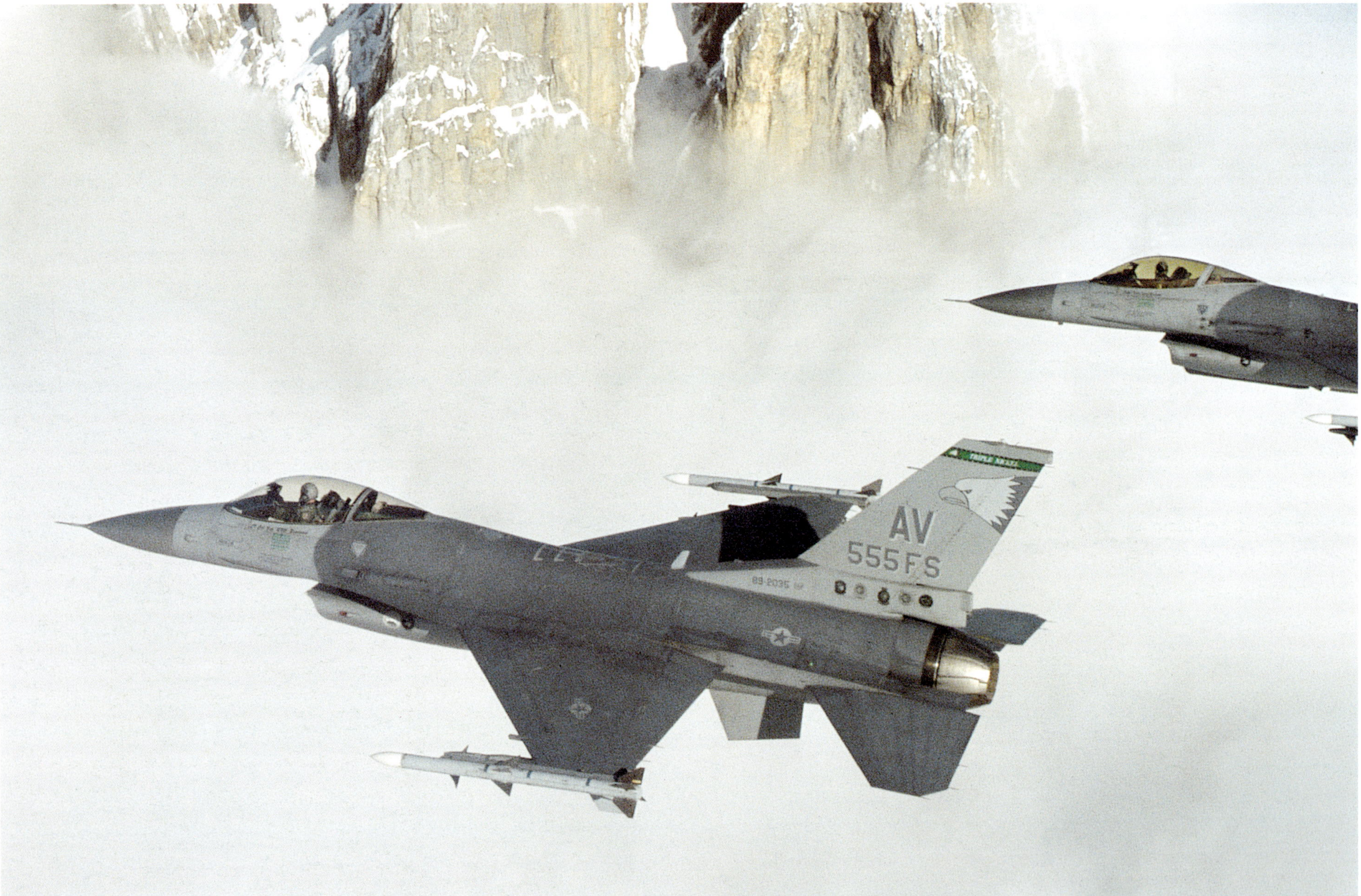

Above: The Alps are the common training environment for the 31st Fighter Wing. Here two Vipers fly close to the high cliffs.

In the page on the left: The Viper in CAP configuration has two wing-tip mounted AIM-120 AMRAAM active-radar-guided missiles and two AIM-9L Sidewinder IR-guided missiles slung under the wings.

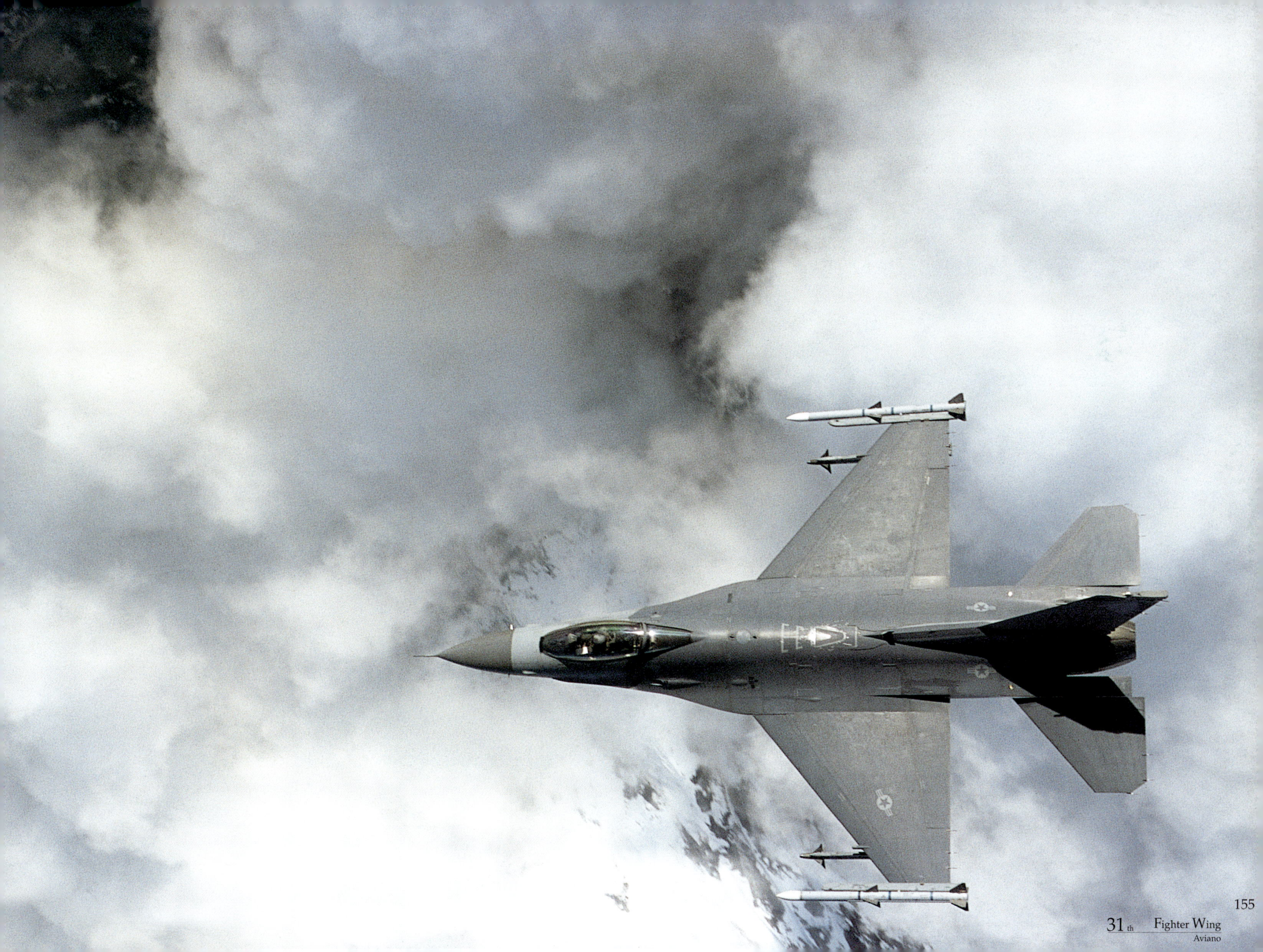

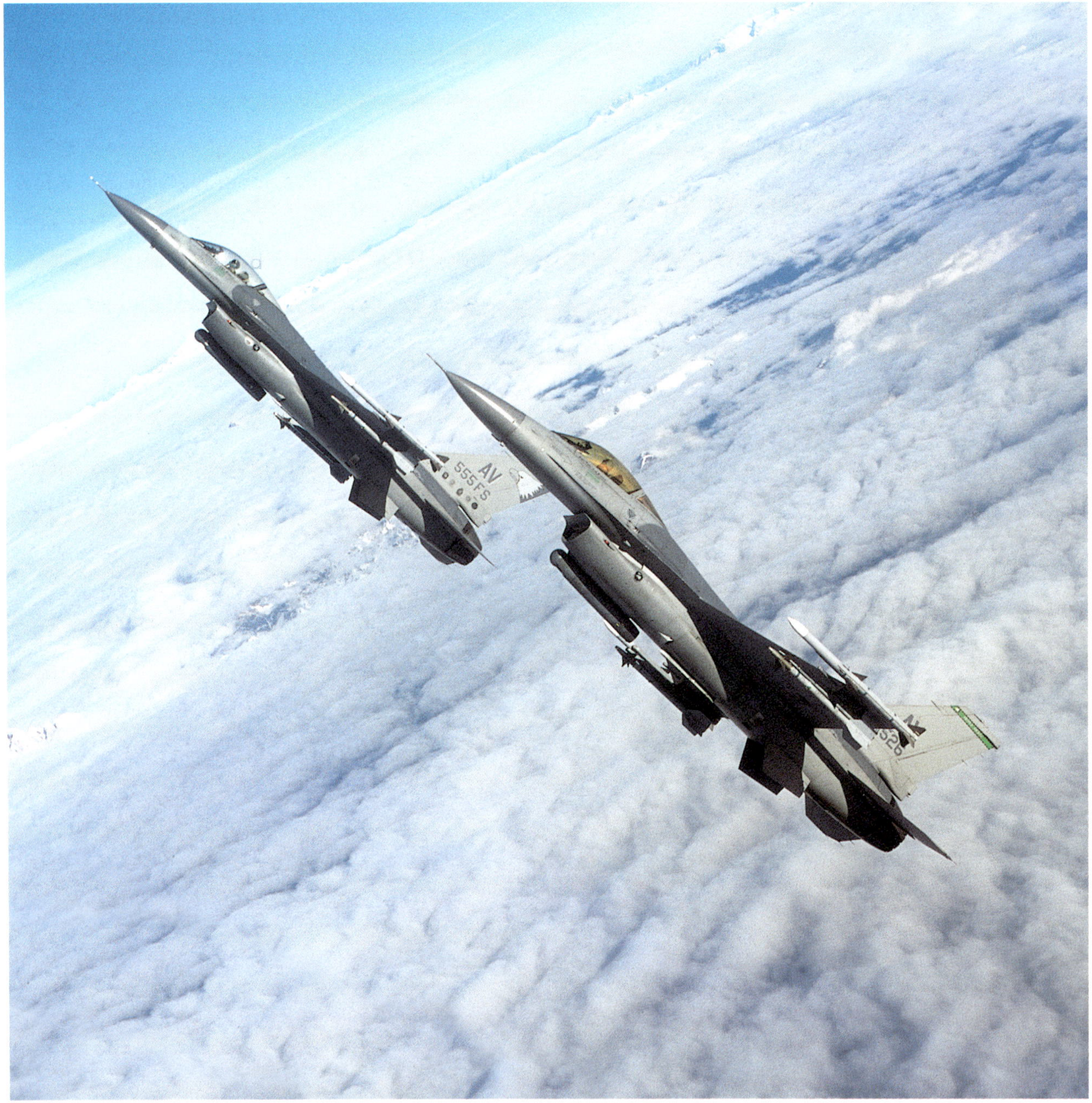

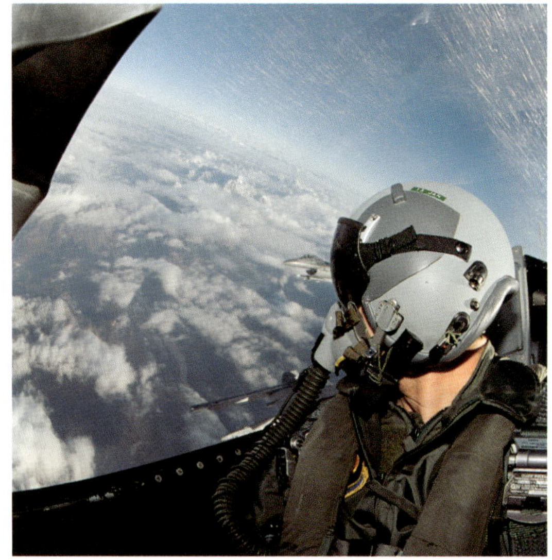

Above
Checking position of two "Triple Nickel" wingmen in flight over the Alps.

On the left
Two 555th FS Vipers climb in vertical under the exhuberant thrust of their afterburning turbofan engines.

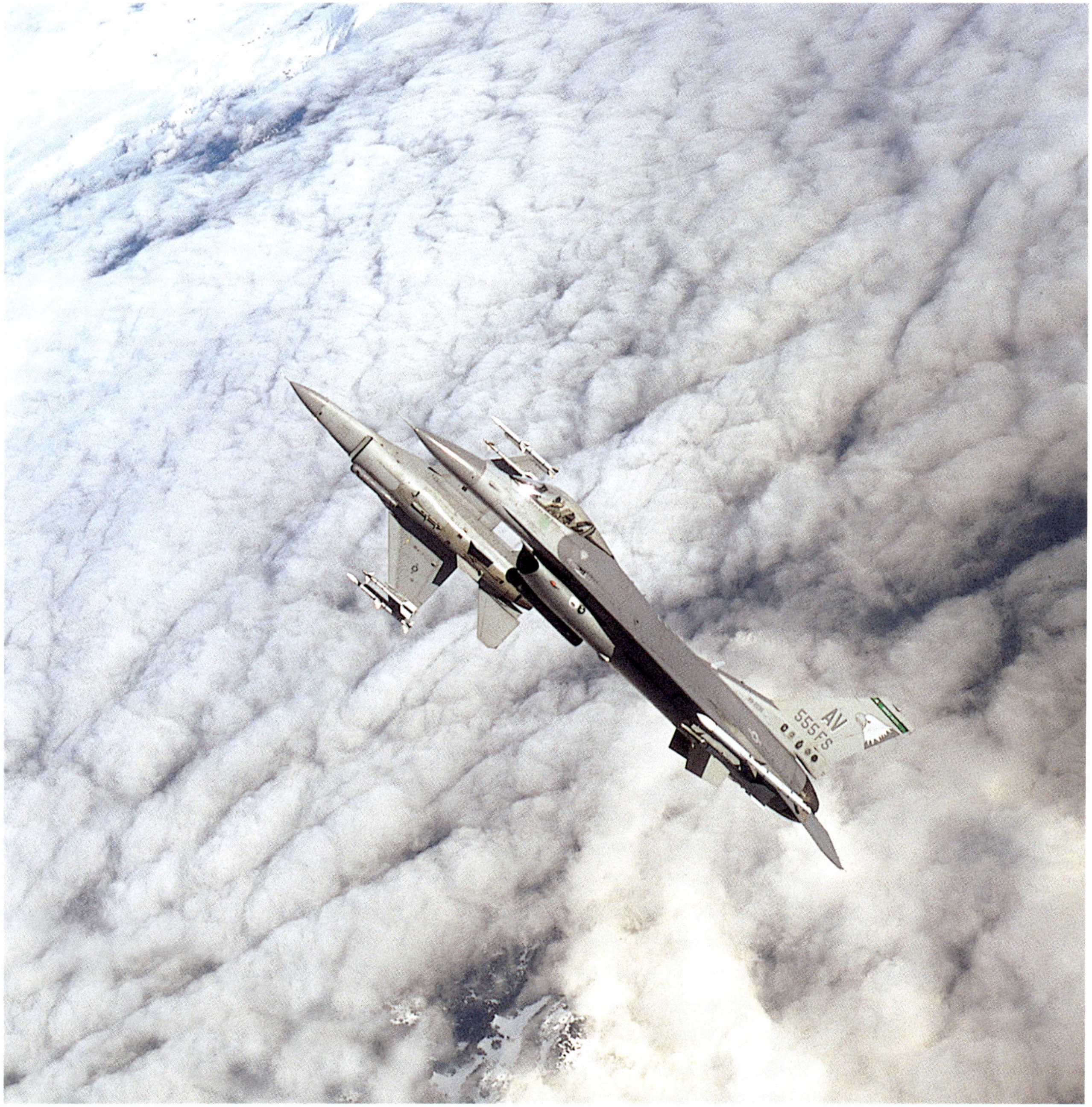

This use of the LANTIRN pod is one of the greater success stories of the air campaign. Our use of NVGs was another success. We had guys who had never flown with NVGs trained up to full combat NVG fliers in a few rides. The NVIS aircraft lighting is fantastic. The mods we made before the war made us some big money in this operation. Whoever came up with the idea for a sequenced infrared strobe should get a medal. With NVGs lights fifty miles away look like they are three miles away. But pilots can pick out their wingmen easily at any distance with the flashing strobe. We were able to employ a lot of NVG four-ship tactics during the war.

We targeted to ensure that bombs tracked on simultaneous strikes. Our internal data modem worked well, too. The ability to transmit targeting data from one platform to another was extremely helpful, especially in the Kosovo engagement zone where we worked several target areas.

O the left
The tactical formation splits
atop the vertical climb.

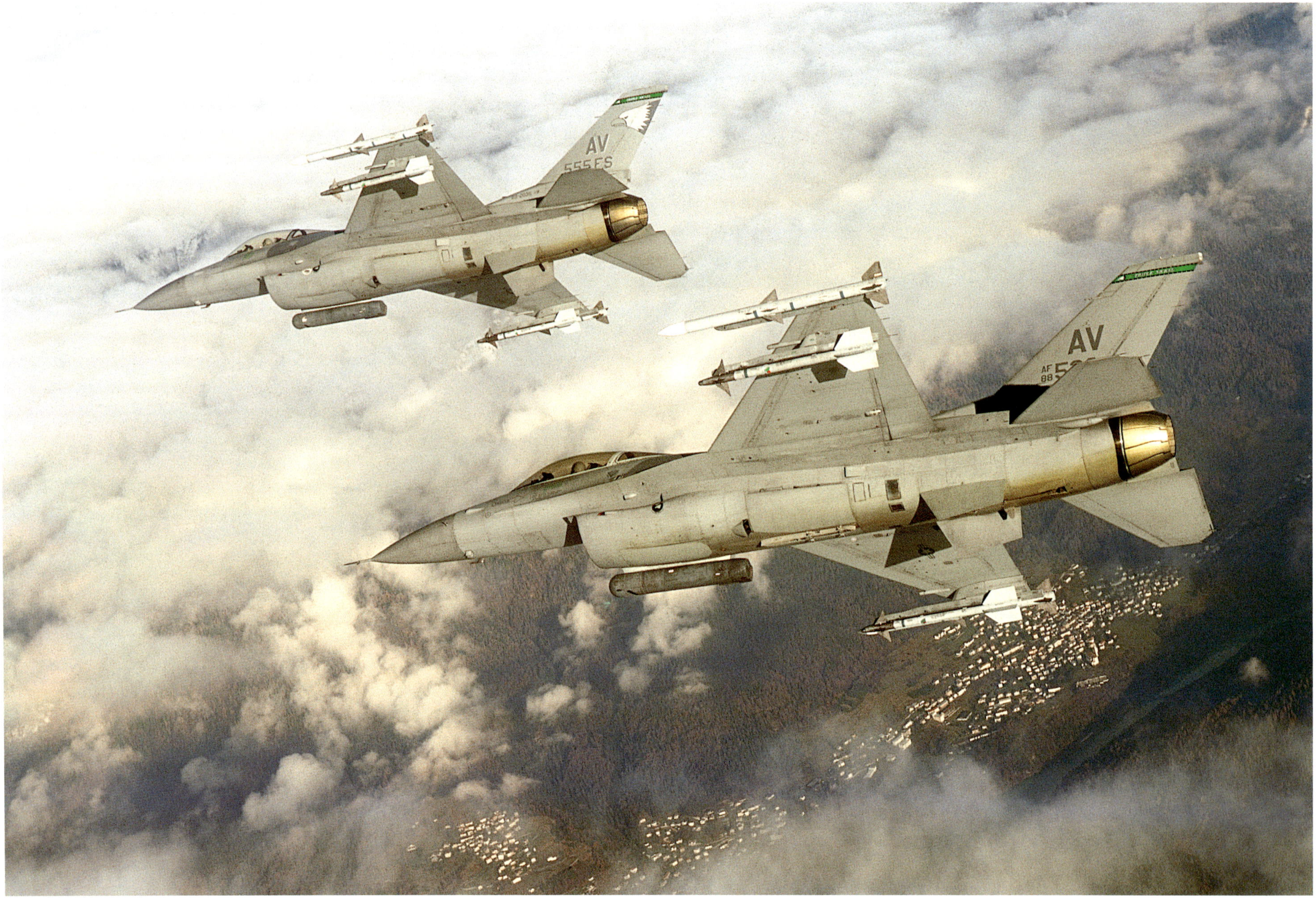
AV
555FS
AV

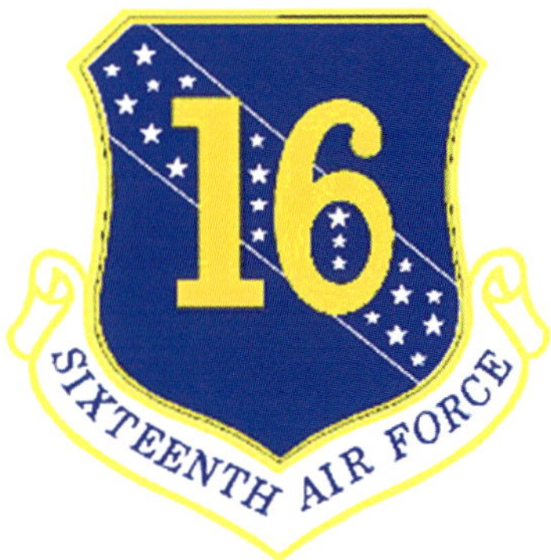

Above
The emblem of the 16th Air Force maintains the "star ribbon" of the Strategic Air Command.

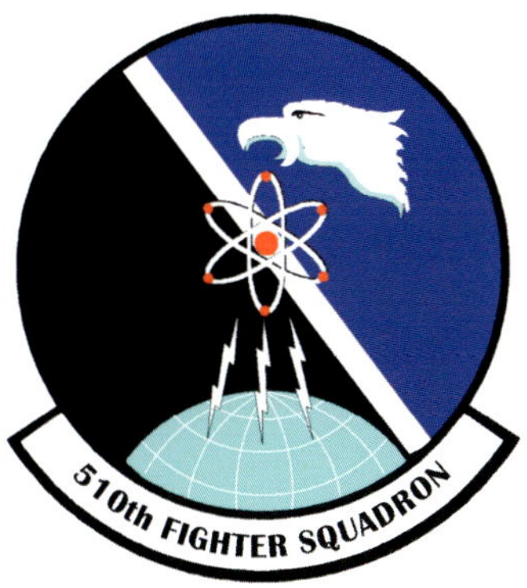

Above
The emblem of the 51th Fighter Squadron, the sister squadron of the "Triple Nickel " in the 31st Fighter Wing.

In the page on the left
Two 41th FW F-16 Block 40 Fighting Falcons equipped with LANTIRN targeting pod.

Sixteenth Air Force

Headquartered at Aviano Air Base, Italy, it is one of two numbered air forces that comprise the U.S. Air Forces in Europe command. It is also the southern air component of the U.S. European Command.

The mission of 16th Air Force is to execute aerospace operations through expeditionary force command and control in support of the U.S. European Command and NATO. The airmen of 16th AF plan, exercise and execute combat air operations, to include initial basing and sustainment of expeditionary forces, in southern Europe and portions of the Middle East and northern Africa. Numbering nearly 11,000 Air Force and civilian members, they conduct peacetime engagement throughout the region from two main operating bases, four support bases and other sites in Spain, France, Germany, Italy, Croatia, Bosnia-Herzegovina, Hungary, Kosovo, the former Yugoslav Republic of Macedonia, Greece, Turkey and Israel.

Three expeditionary wings were activated under 16th AF in 1997. Today those wings exemplify our nation's expeditionary aerospace force through support of Balkans peace enforcement and Iraq airpace control. The 16th AF area of interest also contains waterway choke points to half of the world's shipping, NATO's southern lines of communications, key producers of the world's oil supply, and the crossroads of Islam and Christianity.

Assigned assets include the F-16 Fighting Falcon, with contingency support to the F-15 Eagle, KC-135 Stratotanker, E-3B Sentry (AWACS), U-2 Dragonlady, MC-130P Combat Shadow, HH-60 Pave Hawk, RQ-1A Predator and a full complement of conventional weapons.

16th Air Force operates two main operating bases: Aviano Air Base, Italy, home to the 31st Fighter Wing; and Incirlik AB, Turkey, home to the 39th Wing. These two wings operate support bases in Spain, Italy, Greece and Turkey. In late 1990as the 16th AF was the first Air Force organization to fully employ the expeditionary wing concept. The 16th Air & Space Expeditionary Task Force, a dual role for 16th Air Force, operates three expeditionary wings: the 31st Air Expeditionary Wing, Aviano Air Base, Italy; the 39th Air and Space Expeditionary Wing, Incirlik Air Base, Turkey; and the 16th Air Expeditionary Wing, headquartered at Aviano Air Base, Italy.

16th Air Force history begins in 1954, in Spain, with bases near Madrid, Sevilla and Zaragoza. In 1966 16th AF was reassigned to U.S. Air Forces in Europe. The 401st Tactical Fighter Wing, with three squadrons of F-100Ds, moved from the United States to Torrejon AB, Spain. The wing later converted to F-4 Phantoms, and in 1983, to F-16 Fighting Falcons. The 401st FW moved from Spain to Aviano AB, Italy, in May 1992 and was re-designated as the 31st FW in April 1994.

Author and photographer:	Giampaolo Agostinelli
Graphic designer:	Stefano Fogagnolo
Publisher:	AWD S.r.l. - Italy
Preface:	John Hoffman; Gary Gault; Lt.Gen. Paul K. Weaver, Jr., Former Director, Air National Guard
Text:	Giampaolo Agostinelli; Andrea Artoni; Gary Gault; Maj. Donald Arias; David Somdahl; Dr. Charles J. Gross; Lt. Col. Rod Rodriguez; Capt. Mark Scorsone; Lt.Col. Ryan Orian; Maj. Robert Swertfager; Call Sign "Mule"; Brg. Gen. Mike Worden; Eric Hehs
Editing:	Andrea Artoni
Print separation:	Spinnaker S.r.l. - Italy; Selekta - Italy - Platten Kopie srl - Italy.
Printing:	Pirovano S.r.l. - Italy
Edition:	Second (2003)